Copyright © 2017 Amir M. Mirzendehdel, Krishnan Suresh

First Edition, October 2017

Preface

Origins of this Text

This text has evolved from fundamental research on topology optimization, carried out at the University of Wisconsin, Madison, over the last 8 years or so. It all started with a new topology optimization method called *Pareto*, developed in 2009, under a research grant from National Science Foundation. Pareto was initially implemented using MATLAB, to solve simple 2-D structural problems. It was observed that Pareto was significantly faster than existing topology optimization methods, and could easily handle a variety of constraints.

Motivated by these results, Pareto was re-implemented using C++ for solving large-scale 3-D structural problems. As the research continued, several PhD students, working in Prof. Suresh's laboratory, extended Pareto to address thermo-elastic problems, multi-load problems, multi-body problems, (generic) manufacturing constraints, and more recently, additive manufacturing constraints.

Further, to meet the growing need for topology optimization software, Pareto was made publicly available in 2014 through its cloud implementation at *www.cloudtopopt.com*. Students, engineers and teachers around the world started using Pareto extensively. This, in turn, created a need for a hands-on introduction to topology optimization ... leading to this textbook.

Targeted Audience

The primary audience for this text include senior undergraduate students, first year graduate students and practicing engineers. Given this wide audience, no prior background in topology optimization is assumed; a working knowledge of finite element analysis (FEA) is however helpful. The primary objectives of text are to introduce the readers to topology optimization terminology, discuss and illustrate various sensitivity analysis techniques (that form the backbone of any topology optimization method), provide numerous examples and case-studies to illustrate the merits of topology optimization. While Pareto is used in this text to illustrate the main concepts, the reader can use any topology optimization software that is capable of handling the problems posed in this text.

Pareto Software

Pareto is now as capable as, if not more capable than, many of the commercial topology optimization methods. After years of development at the University of Wisconsin, Madison, it is now exclusively licensed to, and commercially available through, SciArt, LLC (*www.sciartsoft.com*). Readers can request for trial or academic licenses by contacting *support@sciartsoft.com*. Pareto is commercially available both as a stand-alone software for Windows and MacOS, and also as an add-in to SolidWorks (and in the near future, to Creo).

Acknowledgements

Several graduate students and visiting scholars have contributed both directly and indirectly to this textbook. The authors would like to acknowledge Dr. Shiguang Deng, Dr. Josh Danczyk, Dr. Praveen Yadav, Dr. Meisam Takalloozadeh, Dr. Vikalp Mishra, Dr. Chaman S. Verma, Dr. Inna Turevsky, Dr. Kavous Jorabchi, Bian Xiang, Tej Kumar, Buzz Rankouhi, Aaditya Chandrasekhar, Alireza H. Taheri, and Anirudh Krishnakumar. The authors would also like to thank National Science Foundation, Department of Energy, University of Wisconsin Graduate School, Wisconsin Alumni Research Foundation, Sandia National Lab, and Autodesk for their research support.

Amir M. Mirzendehdel	Krishnan Suresh
Postdoctoral Researcher	Philip and Jean Meyer Professor
Amir.Mirzendehdel@parc.com	ksuresh@wisc.edu
www.parc.com	www.ersl.wisc.edu
System Sciences Laboratory	Department of Mechanical Engineering
Palo Alto Research Center	University of Wisconsin, Madison

Contents

1	**Introduction**	9
1.1	**Motivation**	9
1.2	**State of the art**	11
1.3	**TO and additive manufacturing**	12
1.3.1	Support Structures	12
1.3.2	Multi-material	13
1.3.3	Multi-scale	14
1.4	**Objectives of this text**	14
2	**Finite Element Analysis**	17
2.1	**Meshing**	18
2.2	**Stiffness Matrix**	22
2.3	**Linear Solvers**	22
2.3.1	Direct FEA solvers	22
2.3.2	Efficient Iterative FEA solvers	23
3	**Shape Optimization**	25
3.1	**Random Search**	26
3.2	**Shape Optimization Problems**	27
3.2.1	First Order Methods of Optimization	28
3.2.2	Direct Finite Difference Sensitivity of Compliance	30

3.2.3	Indirect Finite Difference Sensitivity of Compliance	30
3.2.4	Scaling for Numerical Robustness	31

4 Topology Optimization 33

4.1 **Density-Based Methods** — **34**
4.1.1 Sensitivity Analysis 35
4.1.2 Algorithm 36

4.2 **Level-set Methods** — **37**
4.2.1 Topological Sensitivity 37
4.2.2 Adjoint Field 39
4.2.3 Pareto Method 39
4.2.4 Discussion 41

5 Getting Started with Pareto 45

5.1 **Pareto** — **46**
5.1.1 ParetoWin 46
5.1.2 ParetoMac 48
5.1.3 ParetoWorks 49
5.1.4 ParetoCloud 49

5.2 **Examples** — **50**

5.3 **Exercises** — **60**

6 Modeling 63

6.1 **Table Design** — **63**
6.2 **Knuckle Design** — **68**
6.3 **Spindle Mount Design** — **72**
6.4 **Exercises** — **77**

7 Performance Objectives & Constraints 79

7.1 **Performance Constraints** — **79**
7.2 **Stress Minimization** — **88**
7.3 **Eigenvalue Maximization** — **92**
7.4 **Additional Examples** — **95**
7.5 **Conclusions** — **101**
7.6 **Exercises** — **102**

8 Design & Manufacturing Constraints 105

8.1 **Draw-Direction Constraint** — **106**
8.2 **Through-Cut Constraint** — **112**
8.3 **Retaining Surfaces** — **115**

8.4	Cyclic Symmetry	119
8.5	Minimum Feature Size	122
8.6	Exercises	126

9 Non-Uniqueness of Designs 129

9.1	Mesh Density	130
9.2	Minimum Feature Size	131
9.3	Step Size	132
9.4	Cyclic Symmetry	133
9.5	Exercises	134

10 Multi-Load Optimization 137

10.1	Multi-Load Strategies	138
10.2	Design of a Crank Arm	140
10.3	FEA on a Multi-Load Problem	142
10.4	Optimizing a Multi-Load Problem	143
10.5	Post-Optimize FEA	144
10.6	Exercises	145

11 Multi-Body & Multi-Material TO 149

11.1	Multi-Body Optimization	149
11.2	Multi-Material Optimization	154
11.2.1	Volume vs. Mass	155
11.2.2	Multi-Objective Formulation	156
11.2.3	Algorithm	157
11.2.4	Initial Material Distribution	162
11.2.5	Numerical Validations	163
11.3	Exercises	171
11.4	References	172

12 Gravity Loading 175

12.1	Body Forces	175
12.2	Examples	176
12.3	Exercises	180

13 Additive Manufacturing 181

13.1	Additive Manufacturing Technologies	182
13.1.1	Material Extrusion	182
13.1.2	Powder Bed Fusion	183
13.1.3	Direct Energy Deposition	184

13.1.4	Vat Photo Polymerization	184
13.1.5	Material Jetting	184
13.1.6	Binder Jetting	185
13.1.7	Sheet Lamination	185
13.2	**Geometry Representation**	**185**
13.2.1	ASCII format	186
13.2.2	Binary format	186
13.2.3	Limitations of STL and Other Options	187
13.3	**Design for Additive Manufacturing**	**188**
13.3.1	Support Structure	188
13.3.2	Anisotropy	194
13.3.3	Surface quality	195
13.3.4	Shrinkage and warping	196
13.3.5	Infill patterns	197
13.4	**Lattice Structures**	**197**
13.4.1	Lattice Unit Cell	198
13.4.2	Example	199
13.5	**Topology Optimization for Additive Manufacturing**	**201**
13.5.1	Support Structure	202
13.6	**References**	**220**
14	**Case Studies**	**221**
14.1	**GE-GrabCAD**	**221**
14.2	**Alcoa-GrabCAD**	**224**
15	**Appendix: Pareto Menus**	**227**
15.1	**Units**	**227**
15.2	**Geometry**	**228**
15.3	**Material**	**228**
15.4	**Loads**	**229**
15.5	**Body Force**	**231**
15.6	**Display**	**231**
15.7	**Finite Element Analysis**	**232**
15.8	**TopOpt Constraints**	**233**
15.9	**Topology Optimization**	**234**
15.10	**TopOpt Results**	**235**
15.11	**Lattice**	**235**
15.12	**Projects**	**236**

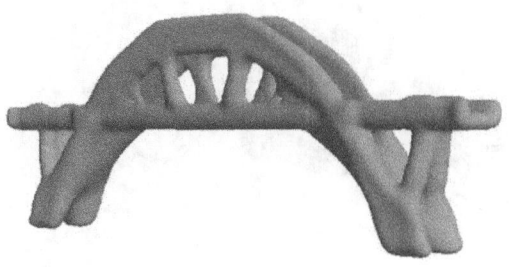 

1. Introduction

Topology optimization (TO) is an exciting method for generating insightful and creative designs. There are several manuscripts and publications that address the theoretical foundations of topology optimization. The objective of this text is to offer a *hands-on* introduction to topology optimization, by juxtaposing theory with examples and case-studies. For supplementary reading, please see:

1. *Homogenization and Structural Topology Optimization: Theory, Practice and Software*, B. Hassani, E. Hinton, 1st edition, Springer, 1999.
2. *Topology Optimization - Theory, Methods, and Applications*, M. P. Bendsoe, O. Sigmund, 2nd Edition Springer, 2004.
3. *Topology Optimization in Engineering Structure Design*, Zhu, J., Gao, T., 1st edition, Elsevier, 2016.

1.1 Motivation

For engineering firms to be competitive, they must constantly create innovative designs by reducing cost, while meeting performance goals. As an example, consider the support bracket in Figure 1.1a. Engineers must figure out where and how to remove material, while ensuring that the part will not fail. One of the tools engineers use for predicting failure is finite element analysis (FEA). Through FEA one can obtain, for example, the stress distribution as illustrated in Figure 1.1b.

Using this stress result, engineers typically remove material where the stresses are relatively low. Figure 1.2 illustrates a few designs that an engineer might generate. Unfortunately, this trial-and-error process is not only laborious, it can lead to designs that are sub-optimal. Furthermore, when additional performance and manufacturing constraints are imposed, this manual approach becomes impractical.

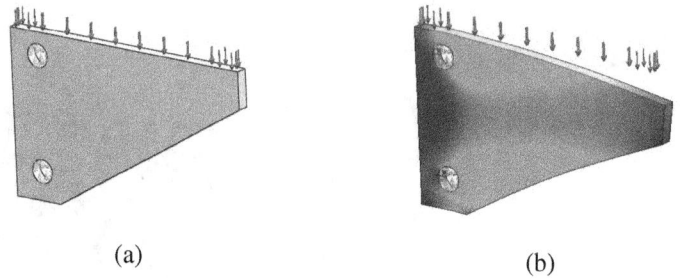

Figure 1.1: (a) A structural problem; the objective is to reduce the weight, (b) FEA can check for failure, and can provides hints where to remove material.

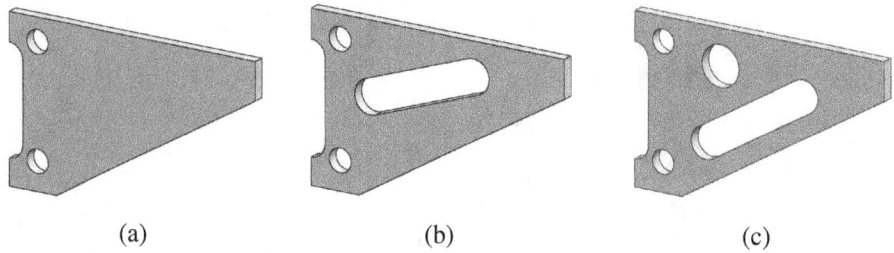

Figure 1.2: Manual generation of light-weight designs can be laborious.

On the other hand, through topology optimization (TO), one can create optimal designs with minimal effort. TO integrates mathematical and physical modeling, with efficient numerical analysis, to generate high-performance light-weight designs. Figure 1.3 illustrates a typical TO workflow that automates the process of material removal, resulting in a highly-optimized design. Observe that the workflow relies on FEA, and on additional concepts such as sensitivity analysis; these concepts are covered later in the text.

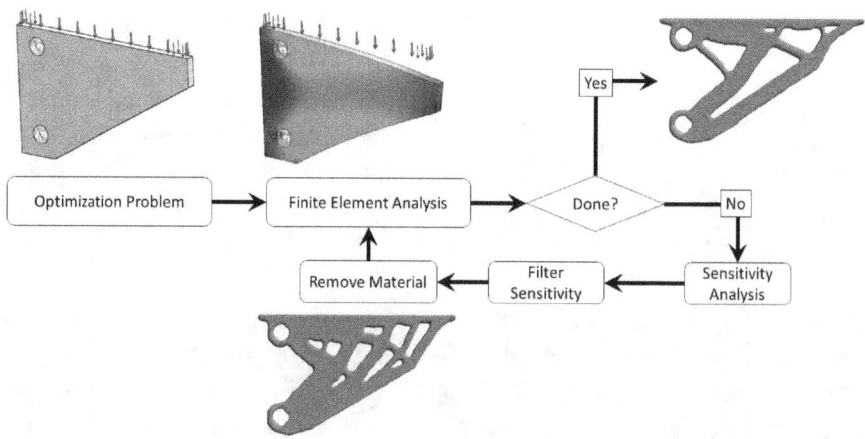

Figure 1.3: Topology optimization work-flow.

1.2 State of the art

Current TO methods are capable of solving multi-load, multi-physics, multi-material, design optimization problems subject to multiple manufacturing and performance constraints. For example, Figure 1.4a illustrates the famous GE-GrabCAD multi-load design problem; the corresponding optimized topology, obtained through TO, is illustrated in Figure 1.4b.

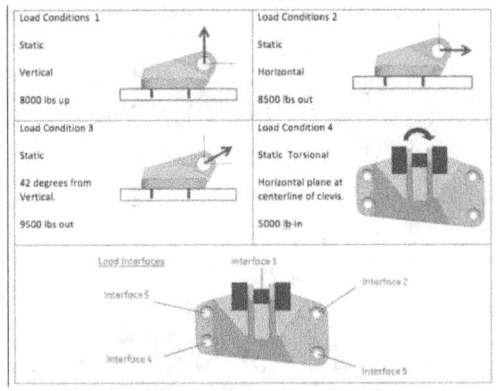
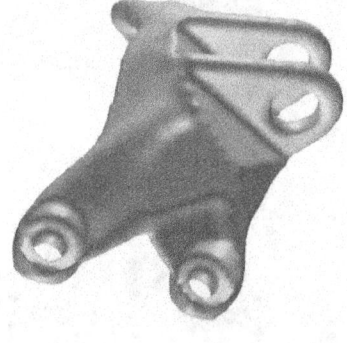

(a) GE-GrabCAD problem   (b) Topology optimized solution

Figure 1.4: TO methods are now capable of solving complex multi-load problems.

Similarly, Figure 1.5a illustrates a structural design problem, that is also subject to a thermal load of 10 degrees Celsius. If the temperature increase is neglected, one would arrive at the optimal topology in Figure 1.5b (incorrect solution). When one considers the thermal load, a different (correct) topology illustrated in Figure 1.5c is obtained.

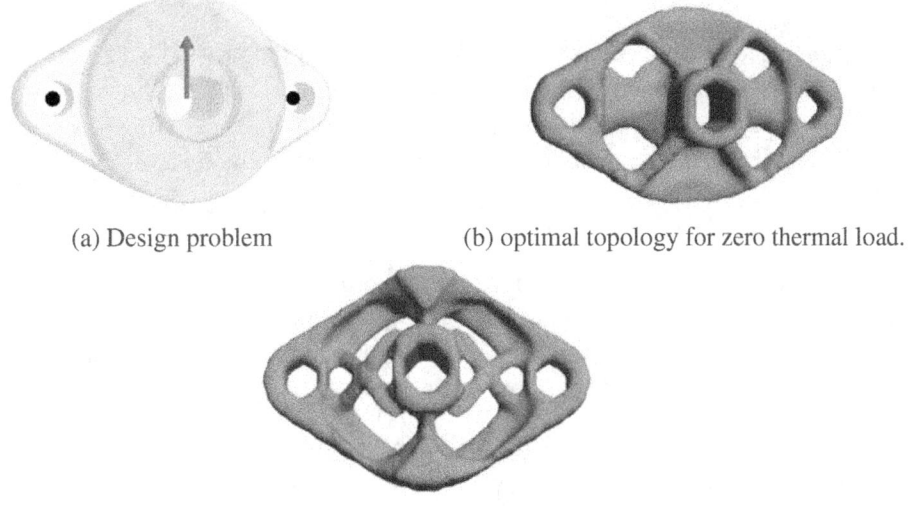

(a) Design problem   (b) optimal topology for zero thermal load.

(c) optimal topology with a temperature increase of 10°C.

Figure 1.5: TO methods are now capable of solving complex multi-physics thermo-elastic problems.

While TO is extremely powerful, designs generated through TO are often organic and geometrically complex. Fabricating these free-form designs through traditional subtractive processes is challenging, if not impossible. Fortunately, one can include manufacturing constraints in TO, as we shall see later in the text. Figure 1.6 illustrates various topologies computed, by imposing manufacturing and design constraints.

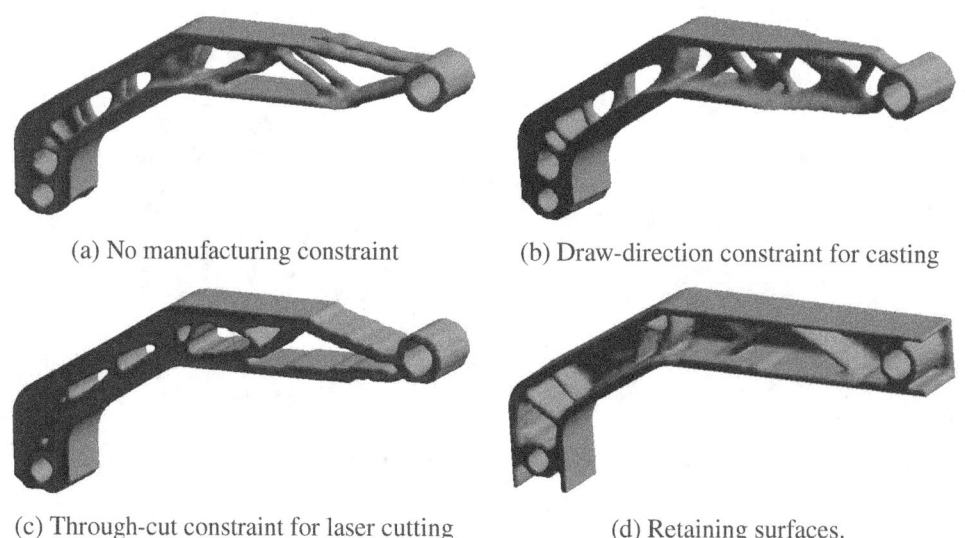

(a) No manufacturing constraint

(b) Draw-direction constraint for casting

(c) Through-cut constraint for laser cutting

(d) Retaining surfaces.

Figure 1.6: Various manufacturing constraints can be included in topology optimization.

1.3 TO and additive manufacturing

The examples above illustrate inclusion of conventional (subtractive) manufacturing constraints. Over the past several years, additive manufacturing (AM) has emerged as a promising alternate to subtractive methods. AM refers to a class of manufacturing processes through which parts are fabricated by material addition. The growing interest in AM stems from its ability to fabricate highly complex parts, with minimal effort.

AM and TO complement each other in that organic and complex designs generated through TO can be easily manufactured through AM. Despite the obvious synergy, there are several challenges that need to be addressed before TO and AM can be seamlessly integrated; these are briefly summarized below, and discussed in detail later in the text.

1.3.1 Support Structures

One of the challenges in AM is the need for extraneous support structures, especially in AM processes such as fused deposition modeling (FDM), selective laser melting (SLM), etc. Figure 1.7 shows a 3D printed GE engine bracket, with additional support structures. Support structures increase material usage, build time, and post-fabrication clean-up time.

1.3 TO and additive manufacturing

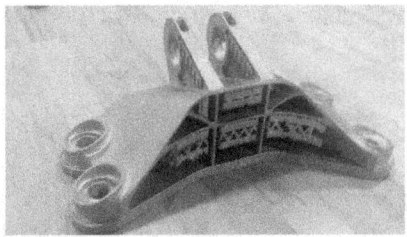

Figure 1.7: Optimized GE engine bracket printed by Arcam (EBM) metal 3D printer at EWI.

Fortunately, one can modify the TO algorithm to reduce support structures as well. For example, Figure 1.8a illustrates a 3D design problem where the objective is to find the optimal topology of 0.7 volume fraction, Figure 1.8b illustrates two different topologies: the first one corresponds to unconstrained compliance minimization, while the second one includes a constraint to reduce support volume by 20%.

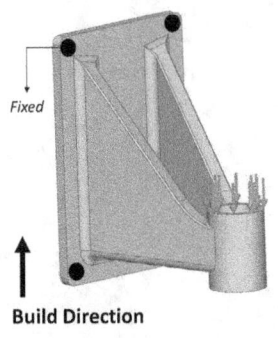

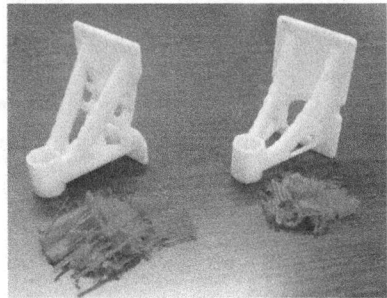

(a) Design problem.

(b) Optimal topologies (left) without and (right) with support structure constraint.

Figure 1.8: Support structures can also be minimized using TO.

1.3.2 Multi-material

There are also additional opportunities that TO opens up that enables us to use multiple materials in AM. For example, consider the mount bracket example of Figure 1.8. But, now we wish to optimize both the topology and material distribution of two materials A and B, whose properties are summarized in Table 11.1.

Table 1.1: Material Properties of A and B.

Material	$E(GPa)$	ν	$\rho(Kg/m^3)$
A	170	0.3	7100
B	70	0.33	2700

Figure 1.9 illustrates the optimized optimized designs using a single material (only A) and both materials (A and B). Although both designs have similar masses, the multi-material

design exhibits superior performance. We will discuss multi-material optimization later in the text.

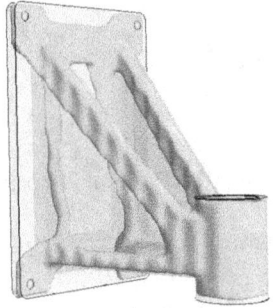 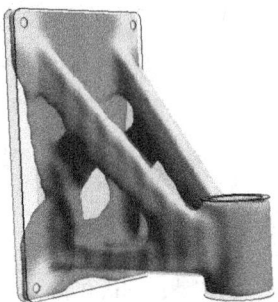

(a) Optimized with only material A.   (b) Optimized with both materials A and B.

Figure 1.9: Both designs have the same mass, while multi-material design is stiffer.

1.3.3 Multi-scale

Another active area of research is the design of microstructures (see below) that exhibit auxetic elastic and thermal properties such as negative Poisson's ratio. These designs can now be created using TO, and manufactured through AM. We will discuss the generation of microstructures and lattice structures as well.

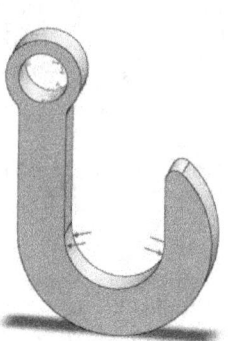 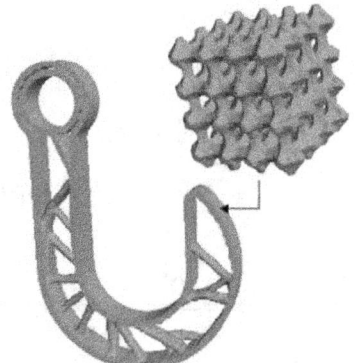

(a) Design problem.   (b) Hierarchical lattice structure.

Figure 1.10: Performance of designs can be improved by optimizing design in macro and micro scales.

1.4 Objectives of this text

As mentioned earlier, the objective of this text is to provide a *hands-on* introduction to topology optimization, while covering the essential mathematics. Specifically, the goals are to:
1. Introduce topology optimization (TO) terminology.
2. Discuss and illustrate various sensitivity analysis techniques for topology optimization.

1.4 Objectives of this text

3. Provide numerous examples and case-studies for readers to appreciate the use of topology optimization.
4. Discuss and illustrate the impact of modeling and constraints on the optimized designs.
5. Finally, encourage readers to confidently use topology optimization in their applications.

The remainder of this book is organized as follows:

- In Chapter 2, we briefly discuss the theory of finite element analysis (FEA), covering concepts such as element types, assembly of stiffness matrices, and linear solvers, within the context of TO.
- In Chapter 3, to differentiate topology optimization from shape optimization, we illustrate the latter through an illustrative example.
- In Chapter 4, we will provide an overview of topology optimization (TO), and discuss two popular TO strategies.
- In Chapter 5, we will introduce a specific 3D cross-platform topology optimization software, namely, *Pareto*, that will be used throughout this text to illustrate various concepts.
- With this background, in Chapter 6, we will study some of pitfalls in modeling for TO, and discuss the designer's role within this context.
- In Chapter 7, we will discuss formulating TO problems considering different performance objectives and constraints and computing corresponding topological sensitivities.
- In Chapter 8, we will consider various design and manufacturing constraints, and study their impact on the optimized designs.
- In Chapter 9, the non-uniqueness of designs obtained through TO is discussed, and illustrated through examples.
- In Chapter 10, we consider multi-load problems, and discuss strategies to address these computationally intensive problems.
- Chapter 11 focuses on optimizing assemblies through multi-body optimization, and optimizing designs using multiple materials.
- In Chapter 12, we include acceleration loads in TO; examples of acceleration loads include gravitational and centrifugal loads.
- In Chapter 13, we give a brief introduction to various additive manufacturing (AM) technologies and discuss the synergy between TO and AM. We will also explain some of the design guidelines for AM, and briefly discuss various lattice structure designs, and contrast this against topology optimization.
- In Chapter 14, we consider several case studies to demonstrate the robustness and capabilities of TO for industrial applications.
- Finally, Chapter 15 serves as a reference for *Pareto*, the primary software used in this text.

2. Finite Element Analysis

Finite element analysis (FEA) is a powerful numerical method to simulate physical phenomena, and it plays a central role in topology optimization (TO). While any generic FEA engine can be used in TO, a highly efficient and customized FEA engine can significantly accelerate TO. In this chapter, we will briefly study how traditional FEA can be modified to speed-up TO.

Towards this end, consider Figure 2.1 that illustrates a typical FEA workflow:

1. One starts with a physical problem defined over a geometry; here a structural problem is illustrated.
2. The geometry is discretized (meshed) into finite elements.
3. Next, for each element, an element stiffness matrix is constructed; these element matrices are then assembled into a global stiffness matrix.
4. The resulting system of equations is then solved to find the unknown field.
5. Finally, post-processing is carried out to compute other derived fields.

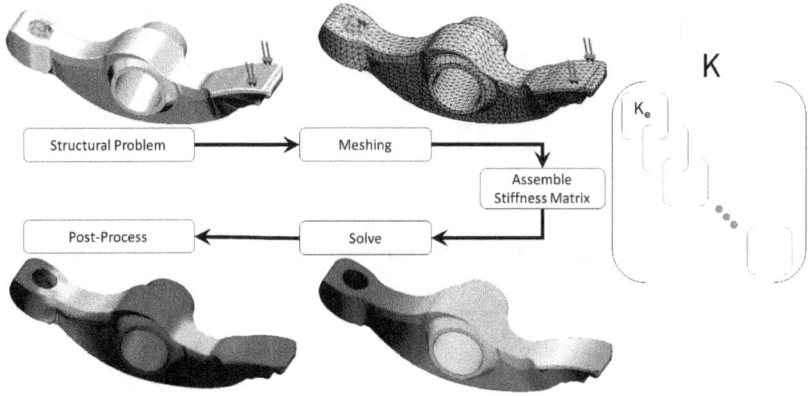

Figure 2.1: Finite element analysis workflow.

It is beyond the scope of this text to delve deep into FEA; for a thorough treatment of FEA concepts, please see:

1. *The Finite Element Method: its basis and fundamentals*, O. C Zienkiewicz, R. L. Taylor, and J. Zhu, Oxford, 2006.
2. *An Introduction to the Finite Element Method*, J. Reddy, 3th edition, McGraw-Hill Education, 2005.
3. *Concepts and Applications of Finite Element Analysis*, R. D. Cook, D. S. Malkus, M. E. Plesha, and R. J. Witt, 4th Edition, Wiley, 2001.

2.1 Meshing

As noted in Figure 2.1, a critical step in FEA is *meshing*, i.e., the discretization of geometry into simpler geometries called *finite elements*. These elements are often triangular or quadrilateral in 2D, and tetrahedral or hexahedral in 3D. Table 2.1 summarizes the most popular types of elements. Higher order elements, such as the quadratic triangle, lead to higher accuracy, but at an increased computational cost. In TO, linear quadrilateral elements are often used in 2D, while linear hexahedral elements are used in 3D; they offer an excellent trade off between accuracy and computational-cost.

Table 2.1: Mesh types

Element	Name	Dimension	# Nodes	Interpolation
	Triangle	2D	3	Linear
	Quadrilateral	2D	4	Linear
	Quadratic triangle	2D	6	Quadratic
	Quadratic quadrilateral	2D	8	Quadratic

continued ...

2.1 Meshing

...continued

Element	Name	Dimension	# Nodes	Interpolation
	Tetrahedral	3D	4	Linear
	Hexahedral	3D	8	Linear
	Quadratic tetrahedral	3D	10	Quadratic
	Quadratic hexahedral	2D	20	Quadratic

In classic FEA, the mesh typically conforms to the boundary of the domain; Figure 2.2(a), for example, illustrates a conforming tetrahedral mesh.

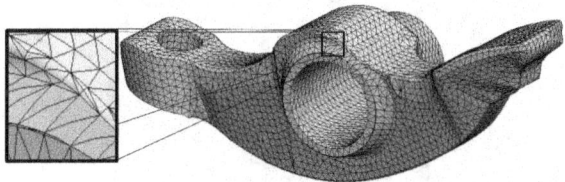

(a) Conforming tetrahedral mesh.

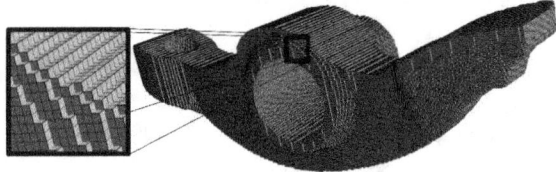

(b) Non-conforming voxel mesh.

Figure 2.2: Comparison between conforming and non-conforming meshes.

Such conforming meshes result in high accuracy (compared to a non-conforming mesh), and are especially desirable during the final stages of design. However, they are hard to generate, and hard to justify in TO. In TO, the boundary evolves during optimization, defeating the very purpose of generating a conforming mesh. Further, repeated construction of a conforming meshes (during optimization) can be very expensive. Instead, the most common type of mesh used in TO is a structured non-conforming quadrilateral mesh in 2D, and a structured non-conforming *voxel* (**vo**lumetric pi**xel**) mesh, comprising of identical linear hexahedral elements in 3D. Figure 2.2(b) illustrates a non-conforming voxel mesh. Structured meshes are easy to generate and robust with respect to geometric complexity. Further, one can exploit the uniformity of the mesh to accelerate FEA through assembly-free methods. For example, given an identical number of elements, one can achieve a 5X FEA speed-up using a 3D voxel-mesh, compared to an unstructured hexahedral mesh.

To understand the process of voxelization, let us consider a 2D case, for simplicity; extension to 3D is straight forward. Consider the 2D geometry of Figure 2.3a; the first step in voxelization is to compute its bounding box (Figure 2.3b). The first step after finding the

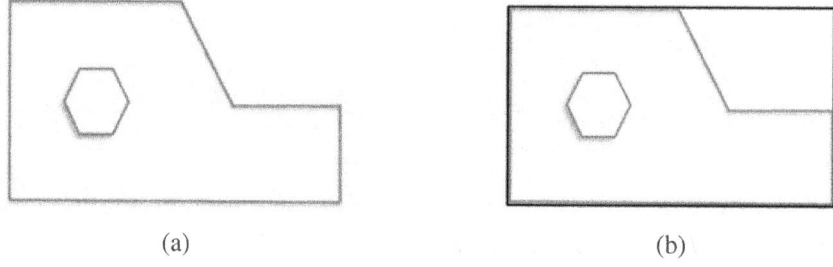

Figure 2.3: (a) Geometry and (b) Bounding box

bounding box is to create a background grid, given a user-defined grid size h_x and h_y.

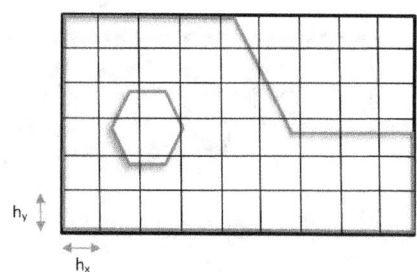

Figure 2.4: Background grid.

Next, we cast equally-spaced rays in y direction, and find the intersection points with the boundary (Figure 2.5a). After gathering all intersection points for each ray, we sort the nodes with respect to their y coordinates. Observe that for a valid sketch, there are always an even number of intersection points (unless the ray happens to tangentially touch the sketch). Then we create line segments starting from the first intersection point and the second one

(Figure 2.5b). A node in the grid is within the geometry if it lies within any line segment. Using this classification, the grid elements are classified as being in or out (Figure 2.5c). One can easily improve the accuracy by increasing the voxel density (Figure 2.5d).

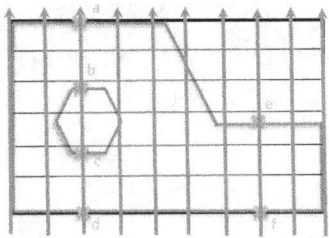

(a) Ray tracing and boundary intersections.

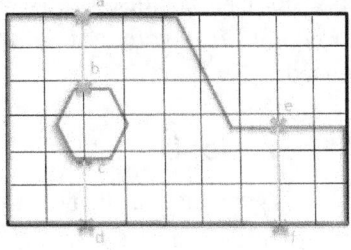

(b) Segments inside the geometry.

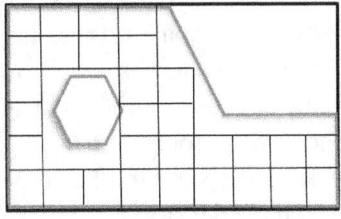

(c) Discard outside elements.

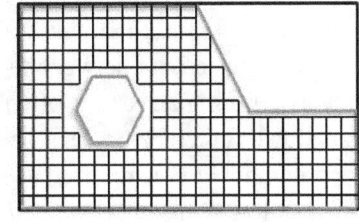

(d) Improve resolution.

Figure 2.5: An overview of the voxelization process.

In 3D, the boundary consists of triangles (see Figure 2.6a), and rays equally-spaced in x and y, are cast in the z direction. The voxelization of the geometry in Figure 2.6a is illustrated in Figure 2.6b. The voxelized model has about 300,000 voxels, and can capture the geometry with great detail.

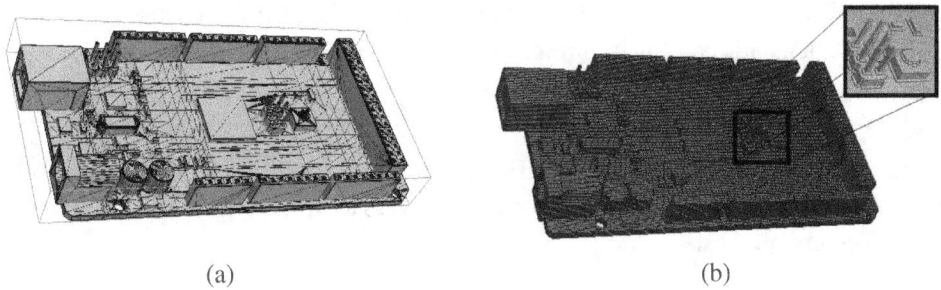

(a)   (b)

Figure 2.6: Arduino MEGA 2560 (a) triangulated surface and bounding box and (b) Voxelized.

> For a deeper discussion on voxelization, please see:
> - Wang, Sidney W., and Arie E. Kaufman. *Volume sampled voxelization of geometric primitives*, Visualization, IEEE Conference, 1993.

While the voxelization process is efficient and robust, a typical argument against the use of voxel meshes in FEA is that they introduce stress singularities. While this is true, such stress singularities are usually acceptable in TO for the following reasons: (1) TO is used during the early stages of design, and (2) one can circumvent stress singularities via global stress norms (to be discussed). Thus, *within the context of TO, the benefits of a voxel mesh outweigh its disadvantages*. However, a conforming FEA of the optimized design is recommended.

2.2 Stiffness Matrix

Once a mesh is generated, a stiffness matrix must be computed for each element. However, observe that, for a voxel mesh, since all elements are identical, only one element stiffness matrix needs to be constructed ... yet another advantage of a such meshes!

The assembly of these element stiffness matrices into a global stiffness matrix **K** is governed by the connectivity between the elements (please see references above for an explanation). The resulting matrix **K** is often sparse (with few non-zero entries) and symmetric. Similarly, a global force vector **f** is assembled using the element force vectors. The number of rows (or columns) in **K** is referred as the *degrees of freedom* (DOF) for the problem and it can vary from a few thousands, to millions, depending on the density of the mesh. Finally, the stiffness matrix and force vector are used to define a linear system of equations:

$$\mathbf{K}\mathbf{u} = \mathbf{f} \tag{2.1}$$

where **u** is the unknown nodal field distribution.

2.3 Linear Solvers

Once the linear system of Equation 2.1 is formed, one must *efficiently* solve this system. Since TO relies on numerous FEA evaluations, solving Equation 2.1 is often the bottleneck, and therefore demands careful attention.

2.3.1 Direct FEA solvers

A popular class of methods for solving Equation 2.1 are *direct* solvers where the stiffness matrix is *factorized*, for example, using *Cholesky factorization*, i.e., the global stiffness matrix is decomposed as:

$$\mathbf{K} = \mathbf{L}\mathbf{L}^T \tag{2.2}$$

where **L** is a lower triangular matrix. Then, the finite element solution can be easily computed via:

$$\mathbf{u} = \mathbf{L}^{-T}\mathbf{L}^{-1}\mathbf{f} \tag{2.3}$$

Such direct solvers are robust and highly efficient when the stiffness matrix is relatively small, say, when the DOF is less than hundred thousand. For large systems of equations,

2.3 Linear Solvers

direct solvers are not often the best choice. Further, direct solvers are also memory intensive. For instance, for a system with one million DOF, about one Gigabyte (GB) of memory is required to store the stiffness matrix, but factorization requires an additional 10-20 GB of memory. Thus direct solvers are both CPU and memory intensive.

 Below are two textbooks on matrix algebra that comprehensively explain the theory behind different factorization schemes:
  1. *Numerical Linear Algebra*, L. N. Trefethen and D. B. III, SIAM: Society for Industrial and Applied Mathematics, 1997.
  2. *Matrix Computations*, G. H. Golub and C. F. V. Loan, Fourth edition, Johns Hopkins University Press, 2012.

2.3.2 Efficient Iterative FEA solvers

For large systems of equations, iterative solvers are typically preferred. There are several iterative solvers; *conjugate gradient* (CG), for example, is a popular iterative method for positive definite systems.

To iteratively solve $\mathbf{Ku} = \mathbf{f}$ via CG, we begin with an initial guess $\mathbf{u}_0$ and follow the algorithm below:

Algorithm 2.3.1 Linear Conjugate Gradient
1. $\mathbf{r}_0 = \mathbf{f} - \mathbf{Ku}_0$
2. $\mathbf{p}_0 = \mathbf{r}_0$
3. $k = 0$
4. Do
5. $\alpha_k = \dfrac{\mathbf{r}_k^T \mathbf{r}_k}{\mathbf{p}_k^T \mathbf{Kp}_k}$
6. $\mathbf{u}_{k+1} = \mathbf{u}_k + \alpha_k \mathbf{p}_k$
7. $\mathbf{r}_{k+1} = \mathbf{r}_k - \alpha_k \mathbf{Kp}_k$
8. if $\mathbf{r}_{k+1} < \varepsilon$ terminate with solution $\mathbf{u}_{k+1}$
9. $\beta_k = \dfrac{\mathbf{r}_{k+1}^T \mathbf{r}_{k+1}}{\mathbf{r}_k^T \mathbf{r}_k}$
10. $\mathbf{p}_{k+1} = \mathbf{r}_{k+1} + \beta_k \mathbf{p}_k$
11. $k = k + 1$
12. end Do

Various preonditioners are typically used to accelerate the convergence of CG. Further *matrix-free schemes*, that do not assemble the global matrix, can be used to reduce memory bottlenecks. Finally, iterative solvers are highly amenable to parallel computing. These characteristics make iterative solvers the typical choice in TO.

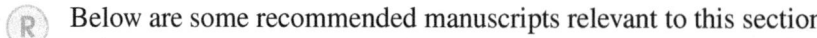 Below are some recommended manuscripts relevant to this section:
1. *An introduction to the conjugate gradient method without the agonizing pain*, J. R. Shewchuk, Carnegie-Mellon University. Department of Computer Science, 1994.
2. *Iterative Methods for Sparse Linear Systems: Second Edition*, Y. Saad, SIAM, 2003.

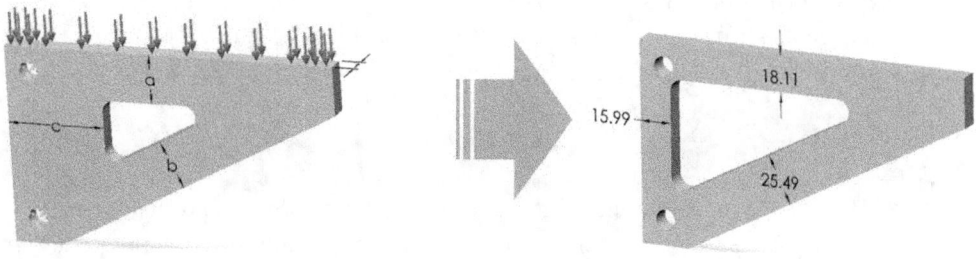

3. Shape Optimization

In this chapter, we shal study how shape optimization problems can be posed. Shape optimization aims to maximize performance (or reduce cost) by modifying design parameters associated with a CAD model. Historically, the theory of shape optimization was developed much before that of topology optimization. However, in a typical design workflow, shape optimization follows topology optimization (TO), i.e., TO is first carried out to arrive at an initial design, followed by shape optimization. In this chapter, we provide an overview of shape optimization since many of the underlying concepts are also applicable to TO.

Consider the support bracket of Figure 3.1; the thickness of the bracket is 10 mm; the underlying material is steel with elastic modulus $E = 2.1e11 N/m^2$ and Poisson's ratio $\nu = 0.28$. The two holes on the left-hand side are clamped, while a pressure of magnitude $5000 KN/m^2$ is applied on the top face.

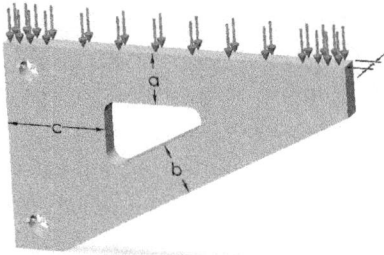

Figure 3.1: Support bracket with shape parameters.

The goal is to optimize the design by modifying the shape parameters a, b, and c, thereby changing the *shape* of the design, without introducing new features into the design.

Specifically, consider the objective of reducing the volume of the part while placing a constraint on its stiffness. A standard measure of stiffness is through its inverse, namely the *compliance*, defined as follows:

$$J = \mathbf{f}^T \mathbf{u} \tag{3.1}$$

where $\mathbf{f}$ and $\mathbf{u}$ are the finite element discretized external force and displacement vectors, respectively. Observe that for a given force, the larger the displacement, the larger the compliance. Thus, the typical objective in shape and topology optimization is to minimize compliance.

3.1 Random Search

A simple but inefficient strategy for finding the optimal shape parmeters is to randomly vary the shape parameters in some acceptable range, and study the compliance of these designs. For instance, for the model in Figure 3.1, we set the following ranges:

$$\begin{aligned} 15mm &\leq a \leq 30mm \\ 10mm &\leq b \leq 30mm \\ 18mm &\leq c \leq 60mm \end{aligned} \tag{3.2}$$

The two extreme designs are illustrated in Figure 3.2.

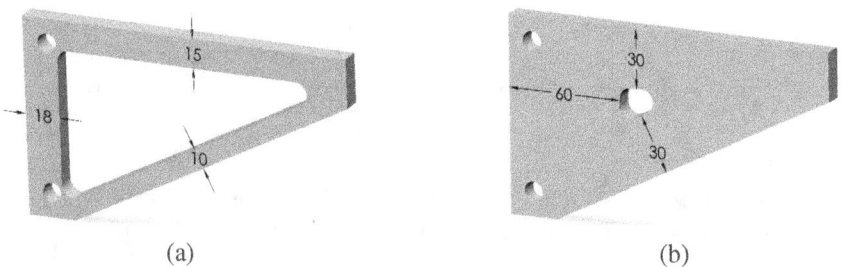

(a)   (b)

Figure 3.2: Designs with (a) lower and (b) upper bound values of design parameters.

Within the above range, one can generate random designs, and for each design, one can to compute the volume and compliance (via FEA). For 200 random samples, Figure 3.3 illustrates the compliance and volume of these designs (relative to the extreme design in Figure 3.2b). Observe the following:
- Among all designs with similar volume, there exists a design with highest stiffness, i.e., least compliance. For example, given a design D_0, there exists a design D_1 that has the same volume as D_0 but is much stiffer, i.e., D_1 is superior to D_0.
- Similarly, among all designs with identical compliance, there exists a design with the lowest volume. For example, given a design D_0, there exists a design D_2 that has the same compliance as D_0 but is of a lower volume, i.e., D_2 is superior to D_0.
- The collection of designs with either the lowest compliance (for a given volume), or lowest volume (for a given compliance) are referred to as *Pareto-optimal* designs. These designs lie on a *Pareto-optimal* curve illustrated in Figure 3.3.

3.2 Shape Optimization Problems

- Designs that lie below the Pareto-optimal curve are infeasible, and designs that lie above the curve are sub-optimal.

The Pareto-optimal curve in Figure 3.3 essentially comprises of all designs that give the best trade-off between compliance and volume. Such optimal curves are referred to as *Pareto frontiers*, named after the Italian economist and engineer, Vilfredo Pareto (1884-1923).

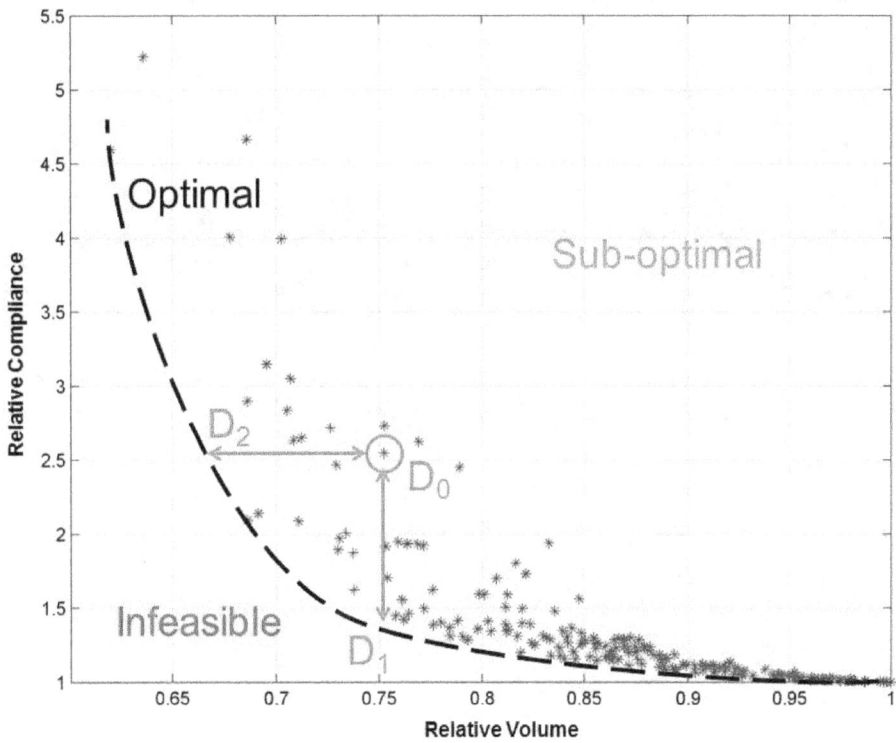

Figure 3.3: A random sample of designs; compliance versus volume.

In general, a solution is Pareto-optimal when further improvement of one criteria worsens at least one other criteria. Such Pareto-optimal designs will play an important role in the remainder of this text.

3.2 Shape Optimization Problems

Based on the above observations, one can now formally pose two meaningful shape optimizations problems:

1. Starting from any design D_0 with volume V_0 and compliance J_0, find the shape parameters such that the compliance is minimized subject to a constraint on the

volume:

$$\begin{aligned}
& \underset{a,b,c}{\text{minimize}} \quad J \\
& V \leq V_0 \\
& a_{min} \leq a \leq a_{max} \\
& b_{min} \leq b \leq b_{max} \\
& c_{min} \leq c \leq c_{max}
\end{aligned} \quad (3.3)$$

Figure 3.4 illustrates a sub-optimal design D_0, and a corresponding Pareto-optimal design D_1.

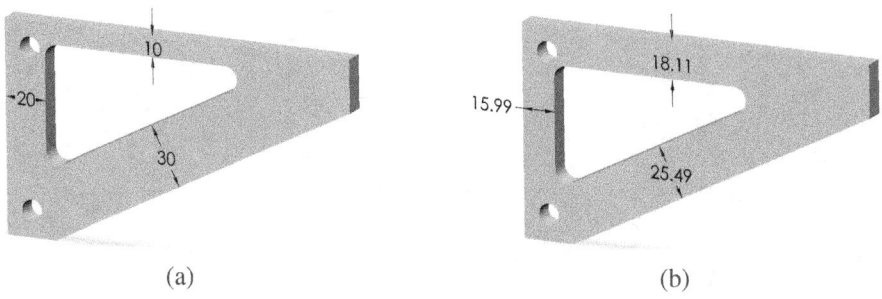

(a)   (b)

Figure 3.4: (a) Initial design (D_0) and (b) shape optimized design at the same volume (D_1).

2. Starting from any design D_0 with volume V_0 and compliance J_0, find the shape parameters such that the volume is minimized subject to a constraint on the compliance:

$$\begin{aligned}
& \underset{a,b,c}{\text{minimize}} \quad V \\
& J \leq J_0 \\
& a_{min} \leq a \leq a_{max} \\
& b_{min} \leq b \leq b_{max} \\
& c_{min} \leq c \leq c_{max}
\end{aligned} \quad (3.4)$$

The remainder of this chapter will focus on the tools needed to solve such shape optimization problems.

3.2.1 First Order Methods of Optimization

There are various methods to solve Equations 3.3 and 3.4. By far, the most popular methods are first order optimization methods. To understand the basic principle behind such methods, let us consider a simple unconstrained optimization problem.

$$\underset{\mathbf{x}}{\text{minimize}} \quad f(\mathbf{x}) \quad (3.5)$$

The classic first order method for solving the above optimization problem is the non-linear conjugate gradient method that proceeds as follows. Starting at an initial guess for solution $\mathbf{x}_0$ and search direction $\mathbf{d}_0 = -\nabla f(\mathbf{x}_0)$, we update solution as follows:

3.2 Shape Optimization Problems

Algorithm 3.2.1 **Non-linear Conjugate Gradient**
1. $k = 0$
2. Do

3. $\mathbf{x}_{k+1} = \mathbf{x}_k + \mathbf{d}_k$

4. $\beta_{k+1} = \dfrac{\nabla f^T(\mathbf{x}_{k+1}) \nabla f(\mathbf{x}_{k+1})}{\nabla f^T(\mathbf{x}_k) \nabla f(\mathbf{x}_k)}$

5. $\mathbf{d}_{k+1} = -\nabla f(\mathbf{x}_{k+1}) + \beta_{k+1} \mathbf{d}_k$

6. if $\|d_{k+1}\| < \varepsilon$ terminate with solution $\mathbf{x}_{k+1}$

7. $k = k + 1$
8. end Do

> **R** A key observation is that one must compute the gradient of the objective $\nabla f(\mathbf{x})$ at each step; the gradient is also referred to as the *sensitivity*. For the remainder of this text, the sensitivity field will play a critical role.

The above algorithm can be extended to handle constraints. For example, consider the following constrained minimization problem:

$$\begin{aligned} & \underset{\mathbf{x}}{\text{minimize}} \, f(\mathbf{x}) \\ & g_i(\mathbf{x}) \leq 0 \quad i = 1, 2, ..., M \\ & h_j(\mathbf{x}) = 0 \quad j = 1, 2, ..., N \end{aligned} \tag{3.6}$$

where g are *inequality* constraints and h are *equality* constraints. Here, one must form the corresponding *Lagrangian* $\mathscr{L}$:

$$\mathscr{L} = f(\mathbf{x}) + \sum_{i=1}^{M} \lambda_i g_i(\mathbf{x}) + \sum_{j=1}^{N} \mu_j h_j(\mathbf{x}) \tag{3.7}$$

where μ_i and λ_i are the Lagrange multipliers. In order to use first order methos, one must typically compute the gradient of the Lagrangian (or the gradient of *augmented Lagrangian*):

$$\nabla_{\mathbf{x}} \mathscr{L} = \nabla f(\mathbf{x}) + \sum_{i=1}^{M} \lambda_i \nabla g_i(\mathbf{x}) + \sum_{j=1}^{N} \mu_j \nabla h_j(\mathbf{x}) \tag{3.8}$$

The gradient can then be used in a modified non-linear conjugate gradient method to find the optimal solution.

> **R** A detailed discussion of augmented Lagrangian method is beyond the scope of this book; see
>   - *Numerical Optimization*, J. Nocedal, S. Wright, Springer Science & Business Media, 2006.

However, the most significant point is that, to solve an optimization problem, one must typically compute the gradient of the objective and constraints, i.e., one must carry out *sensitivity analysis*.

3.2.2 Direct Finite Difference Sensitivity of Compliance

As we have observed, first order methods require the first derivative of the objective and constraints with respect to the design variables. Within the context of shape optimization, we will consider the problem of computing the gradients. Finding the sensitivity of the bound constraints, i.e., the constraints $x_i^{min} \leq x_i \leq x_i^{max}$, is trivial, while the volume (constraint or objective) can also be easily handled through finite difference. The gradient of compliance is however trickier.

Consider computing the sensitivity of compliance via the forward finite difference:

$$\frac{\partial J}{\partial a} = \frac{J(a+\Delta a,b,c) - J(a,b,c)}{\Delta a}$$

$$\frac{\partial J}{\partial b} = \frac{J(b+\Delta a,b,c) - J(a,b,c)}{\Delta b} \quad (3.9)$$

$$\frac{\partial J}{\partial c} = \frac{J(c+\Delta a,b,c) - J(a,b,c)}{\Delta c}$$

In other words, we solve a pair of FEA problems for each parameter, and use the above equation to compute sensitivity. Unfortunately, the direct finite difference method within the context of FEA suffers from serious numerical problems, and will lead to failure of the shape optimization process.

3.2.3 Indirect Finite Difference Sensitivity of Compliance

An alternate method of estimating the compliance sensitivity is based on the following definition:

$$\mathbf{Ku} = \mathbf{f} \quad (3.10)$$

$$J = \mathbf{f}^T \mathbf{u} \quad (3.11)$$

Taking the derivative of both equations with respect to any shape parameter, we have:

$$\mathbf{K'u} + \mathbf{Ku'} = \mathbf{f'} \quad (3.12)$$

$$J' = \left(\mathbf{f}^T\right)' \mathbf{u} + \mathbf{f}^T \mathbf{u'} \quad (3.13)$$

where:

$$(.)' \equiv \frac{\partial}{\partial a} \quad or \quad \frac{\partial}{\partial b} \quad or \quad \frac{\partial}{\partial c} \quad (3.14)$$

3.2 Shape Optimization Problems

In Equation 3.12 $\mathbf{u}'$ is the displacement sensitivity while $\mathbf{K}'$ is the stiffness matrix sensitivity. Since the force does not depend on the shape parameter, the above equations reduce to:

$$\mathbf{K}\mathbf{u}' = -\mathbf{K}'\mathbf{u} \tag{3.15}$$

$$J' = \mathbf{f}^T \mathbf{u}' \tag{3.16}$$

Now consider the problem of computing the stiffness matrix sensitivity $\mathbf{K}'$ with respect to a shape parameter. Specifically, consider the finite difference estimation:

$$\begin{aligned}
\frac{\partial \mathbf{K}}{\partial a} &= \frac{\mathbf{K}(a+\Delta a, b, c) - \mathbf{K}(a,b,c)}{\Delta a} \\
\frac{\partial \mathbf{K}}{\partial b} &= \frac{\mathbf{K}(b+\Delta a, b, c) - \mathbf{K}(a,b,c)}{\Delta b} \\
\frac{\partial \mathbf{K}}{\partial c} &= \frac{\mathbf{K}(c+\Delta a, b, c) - \mathbf{K}(a,b,c)}{\Delta c}
\end{aligned} \tag{3.17}$$

It turns out one can efficiently compute the stiffness matrix sensitivity using mesh morhping methods. Thus, the algorithm to compute compliance sensitivity is as follows:

> **Algorithm 3.2.2 Compliance Shape Sensitivity**
> 1. First, compute the stiffness matrix sensitivity $\mathbf{K}'$ with respect to each shape parameter (this is usually done through mesh morphing strategies)
> 2. Then, compute the displacement sensitivity $\mathbf{u}'$ by solving Equation 3.15; observe that this is similar to solving for the displacement $\mathbf{u}$ in Equation 3.10.
> 3. Finally, estimate the compliance sensitivity via Equation 3.16.

3.2.4 Scaling for Numerical Robustness

From a numerical perspective, additional numerical scaling is essential for robustness. Specifically, given an initial set of values (a_0, b_0, c_0) for the shape parameters, the design variables are scaled by introducing non-dimensional variables:

$$\begin{aligned}
x_1 &= a/a_0 \\
x_2 &= b/b_0 \\
x_3 &= c/c_0
\end{aligned} \tag{3.18}$$

This leads to a set of bounds for the non-dimensional variables:

$$\begin{aligned}
x_1^{min,max} &= a^{min,max}/a_0 \\
x_2^{min,max} &= b^{min,max}/b_0 \\
x_3^{min,max} &= c^{min,max}/c_0
\end{aligned} \tag{3.19}$$

Similarly, using the initial values (J_0, V_0) for the compliance and volume, Equations 3.3 and 3.4 are transformed as follows:

$$\begin{aligned}
&\underset{x_1,x_2,x_3}{\text{minimize}} \quad f = \frac{J}{J_0} \\
&\frac{V}{V_0} - 1 \leq 0 \\
&x_1^{min} \leq x_1 \leq x_1^{max} \\
&x_2^{min} \leq x_2 \leq x_2^{max} \\
&x_3^{min} \leq x_3 \leq x_3^{max}
\end{aligned} \qquad (3.20)$$

and

$$\begin{aligned}
&\underset{x_1,x_2,x_3}{\text{minimize}} \quad f = \frac{V}{V_0} \\
&\frac{J}{J_0} - 1 \leq 0 \\
&x_1^{min} \leq x_1 \leq x_1^{max} \\
&x_2^{min} \leq x_2 \leq x_2^{max} \\
&x_3^{min} \leq x_3 \leq x_3^{max}
\end{aligned} \qquad (3.21)$$

Equations 3.20 and 3.21 are now amenable to numerical optimization. The reader is referred to references cited above for additional details.

4. Topology Optimization

Topology optimization generalizes shape optimization by allowing new features to be introduced into the design. After more than two decades of intensive research, it has established itself as an essential tool for early stage design. A typical topology optimization problem can be stated as follows: *Given a design domain, and a valid (structural, thermal, fluid, ...) problem, find an optimized design within the design domain, that minimizes an objective (such as volume), such that certain performance constraints (such as stiffness, strength,...) and manufacturing constraints (such as minimum feature size) are met.*

Mathematically, a typical structural TO problem can be stated as:

$$\begin{aligned}&\underset{\Omega \subset D}{minimize\, J} \\ &\text{subject to} \\ &V \leq V_0 \\ &\mathbf{K}\mathbf{u} = \mathbf{f}\end{aligned} \qquad (4.1)$$

where:

$$\begin{aligned}&J: \text{Compliance} \\ &V_0: \text{Desired volume fraction} \\ &\Omega: \text{Topology to be computed} \\ &D: \text{Domain within which the topology must lie} \\ &\mathbf{u}: \text{Finite element displacement field} \\ &\mathbf{K}: \text{Finite element stiffness matrix} \\ &\mathbf{f}: \text{External force vector}\end{aligned} \qquad (4.2)$$

The reader may contrast the above formulation against the shape optimization formulation discussed in the previous chapter. In topology optimization, unlike shape optimization, there are no design parameters to be modified. Instead, we must determine where to remove

material by relying on various parameterization of the topology. Currently, the two most common methods to parameterize the topology include:
1. Density-based methods
2. Level-set based methods

These two methods are briefly discussed in the remainder of this chapter. Each method leads to a different *sensitivity analysis*, that forms the back-bone of topology optimization. A good grasp of sensitivity analysis techniques is essential to understanding TO.

4.1 Density-Based Methods

The density-based approach, also known as Solid Isotropic Material with Penalization (SIMP), is the most popular TO method. SIMP has a sound mathematical foundation, is capable of handling various objectives and constraints, and is relatively easy to implement within a finite element environment.

 A simple, yet powerful, Matlab implementation of SIMP is described in:

- *A 99 line topology optimization code written in Matlab*,O. Sigmund, Structural and Multidisciplinary Optimization, 2001. 21(2): p. 120-127.

The basic idea behind FEA based SIMP is that each finite element is associated with a fictitious pseudo-density variable $0 \leq \rho \leq 1$, that essentially parameterizes the topology. The pseudo-densities are optimized to reach the desired objective.

With the introduction of these pseudo-densities, Equation 7.6 reduces to

$$\begin{aligned}
& \underset{\rho_e}{\text{minimize}} \, J \\
& \text{subject to} \\
& \sum_{e=1}^{E} \rho_e v_e - V_0 = 0 \\
& \mathbf{Ku} = \mathbf{f}
\end{aligned} \qquad (4.3)$$

In practice, one must also impose upper and lower limits on the density variables, leading to :

$$\begin{aligned}
& \underset{\{\rho\}}{\text{minimize}} \, J = \mathbf{u}^T \mathbf{K} \mathbf{u} \\
& \sum_{e=1}^{E} \rho_e v_e - V_0 = 0 \\
& \mathbf{Ku} - \mathbf{f} = 0 \\
& \rho_e - 1 \leq 0 \quad e = 1, 2, ..., E \\
& \rho_{min} - \rho_e \leq 0 \quad e = 1, 2, ..., E
\end{aligned} \qquad (4.4)$$

There are several optimization methods that have been proposed for solving the above problem. These include method of moving asymptotes, augmented Lagrangian method, etc. Here, we will discuss the optimality criteria method that is simple to implement.

4.1.1 Sensitivity Analysis

To find the optimal values of ρ_e, one must carry out a sensitivity analysis, i.e., one must find the derivative of the objective (and constraints) with respect to these variables. Towards this end, the Young's modulus is interpolated in SIMP as follows:

$$E_e = (\rho_e)^p E_0 \tag{4.5}$$

where the penalization factor p depends on the dimension of the problem; it is typically assigned a value of 3 in 3D.

To compute the sensitivity of a constrained optimization problem, recall that one must define the Lagrangian; here we define the Lagrangian with both $\mathbf{u}$ and ρ as independent variables

$$\mathscr{L}(\mathbf{u},\boldsymbol{\rho}) = \mathbf{u}^T \mathbf{K} \mathbf{u} + \mu_\alpha \left[\sum_{e=1}^{E} \rho_e v_e - V_0 \right] + \mu_\mathbf{f}^T (\mathbf{K}\mathbf{u} - \mathbf{f}) \tag{4.6}$$

Considering the gradient of the Lagrangian with respect to $\mathbf{u}$, we have:

$$\nabla_\mathbf{u} \mathscr{L}(\mathbf{u},\boldsymbol{\rho}) = 2\mathbf{u}^T \mathbf{K} + \mu_\mathbf{f}^T \mathbf{K} = 0 \tag{4.7}$$

Assuming $\mathbf{K}$ is invertible, we have:

$$\left(2\mathbf{u}^T \mathbf{K} + \mu_\mathbf{f}^T \mathbf{K}\right) \mathbf{K}^{-1} = 0 \tag{4.8}$$

i.e,

$$\mu_\mathbf{f}^T = -2\mathbf{u} \tag{4.9}$$

Substituting Equation 4.9 in Equation 4.6, we have:

$$\mathscr{L}(\mathbf{u},\boldsymbol{\rho}) = \mathbf{u}^T \mathbf{K} \mathbf{u} + \mu_\alpha \left[\sum_{e=1}^{E} \rho_e v_e - V_0 \right] - 2\mathbf{u}^T (\mathbf{K}\mathbf{u} - \mathbf{f}) \tag{4.10}$$

i.e.,:

$$\mathscr{L}(\mathbf{u},\boldsymbol{\rho}) = -\mathbf{u}^T \mathbf{K} \mathbf{u} + 2\mathbf{u}^T \mathbf{f} + \mu_\alpha \left[\sum_{e=1}^{E} \rho_e v_e - V_0 \right] \tag{4.11}$$

Now considering the gradient of the Lagrangian with respect to ρ, we have:

$$\nabla_\mathbf{u} \mathscr{L}(\mathbf{u},\boldsymbol{\rho}) = -\mathbf{u}^T \nabla_\rho \mathbf{K} \mathbf{u} + \mu_\alpha \nabla_\rho \left[\sum_{e=1}^{E} \rho_e v_e - V_0 \right] = 0 \tag{4.12}$$

i.e.,:

$$-\mathbf{u}^T \mathbf{K}_{,\rho_e} \mathbf{u} + \mu_\alpha v_e = 0 \quad \forall e \tag{4.13}$$

Observe that since $\mathbf{K}$ is given by:

$$\mathbf{K}_{,\rho_e} = \sum_{e=1}^{E} \rho_e^p \mathbf{K}_0(e) \tag{4.14}$$

we have:

$$\mathbf{u}^T \mathbf{K}_{,\rho_e} \mathbf{u} = p \mathbf{u}_e^T \rho_e^{p-1} \mathbf{K}_0(e) u_e \tag{4.15}$$

Thus from Equation 4.13:

$$p\rho_e^{p-1} \mathbf{u}_e^T \mathbf{K}_0(e) \mathbf{u}_e = \mu_\alpha v_e \quad \forall e \tag{4.16}$$

4.1.2 Algorithm

The Lagrange multiplier μ_α and corresponding update in the densities are computed via the bi-section method as follows. Observe that at a point other than the stationary point, Equation 4.16 will not be satisfied. Therefore one may define:

$$\mu_{\alpha e} = \frac{p\rho_e^{p-1}\mathbf{u}_e^T\mathbf{K}_0(e)\mathbf{u}_e}{v_e} \quad \forall e \qquad (4.17)$$

Now, the bi-section method proceeds as follows:

> **Algorithm 4.1.1 Element-Based Bi-section algorithm**
> 1. Compute the Lagrange multipliers $\mu_{\alpha e}$ via Equation 4.17
> 2. Next the Lagrange multipliers $\mu_{\alpha e}$ are filtered via the technique summarized in Sigmund's paper mentioned above.
> 3. Then, let $\mu_{\alpha 0} = min(\mu_{\alpha e})$ and $\mu_{\alpha 1} = max(\mu_{\alpha e})$. We now search for the *correct* μ_α between these two extremes, and simultaneously modify the densities as follows:
> 4. Define the midpoint Lagrange multiplier as:
>
> $$\mu_\alpha = \frac{\mu_{\alpha 0} + \mu_{\alpha 1}}{2} \qquad (4.18)$$
>
> 5. Then density is updated as:
>
> $$\rho_e^{i+1} = \rho_e^i \sqrt{\frac{\mu_{\alpha e}}{\mu_\alpha}} \qquad (4.19)$$
>
> with the following constraints:
>
> $$\begin{aligned} \|\rho_e^{i+1} - \rho_e^i\| &\leq \Delta\rho_{max} \\ \rho_{min} \leq \rho_e^{i+1} &\leq 1 \end{aligned} \qquad (4.20)$$
>
> 6. The extremes $\mu_{\alpha 0}$ and $\mu_{\alpha 1}$ are modified as follows:
>
> $$\begin{cases} \mu_{\alpha 0} = \mu_\alpha & \sum_{e=1}^{E} \rho_e > \alpha V_0 \\ \mu_{\alpha 1} = \mu_\alpha & \sum_{e=1}^{E} \rho_e \leq \alpha V_0 \end{cases} \qquad (4.21)$$
>
> 7. Return back to step 3 if convergence is not reached.

In summary, the SIMP algorithm therefore proceeds as follows:

> **Algorithm 4.1.2 SIMP Algorithm**
> 1. The densities are initialized to the specified volume fraction $\rho_e = \alpha$.
> 2. The static problem $Ku = f$ is solved with the current densities ρ_e.
> 3. The densities are updated per the bi-section method described above.
> 4. If convergence in densities is not reached, the algorithm returns to Step 2.

Figure 4.1 illustrates the SIMP method.

4.2 Level-set Methods

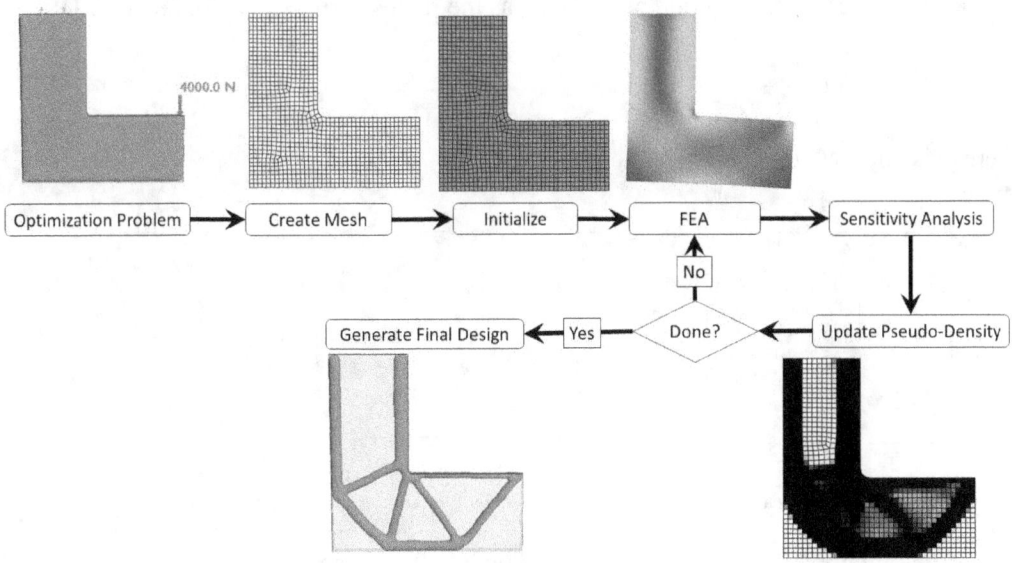

Figure 4.1: SIMP method.

4.2 Level-set Methods

Another popular topology optimization approach is based on level-sets, where material distribution is determined through implicit boundaries or iso-contours. Unlike SIMP, level-set methods create crisp interfaces between material phases. In level-set based methods, the design is optimized by updating the materials interfaces.

> (R) The reader is referred to the following paper for an overview on level-set methods:
>
> - *Level-set methods for structural topology optimization: a review*, N.P. van Dijk, K. Maute, M. Langelaar, F. van Keulen, Structural and Multidisciplinary Optimization, 48, 437–472, 2013.

Early versions of level-set methods were incapable of inserting new holes into the domain. Instead, the domain had to be initialized with a large number of holes, that were allowed to merge during optimization. This limitation was alleviated by the development of new concepts such as *topological sensitivity*, discussed next.

4.2.1 Topological Sensitivity

To illustrate the concept of topological sensitivity, consider the structural problem posed in Figure 4.2a. We shall assume that FEA has been carried, and various displacement and stress fields have been computed. Now consider inserting a *hypothetical* hole of radius r at point p (see Figure 4.2b). One can expect the insertion of the hole to alter the underlying field and various quantities of interest. Topological sensitivity is defined as the ratio of the change in a quantity of interest to the volume of the hole, as the radius shrinks to zero, i.e., in 2-D:

$$\mathcal{T}(p) \equiv \lim_{r \to 0} \frac{\varphi(r;p) - \varphi}{\pi r^2} \qquad (4.22)$$

where φ is any scalar quantity of interest.

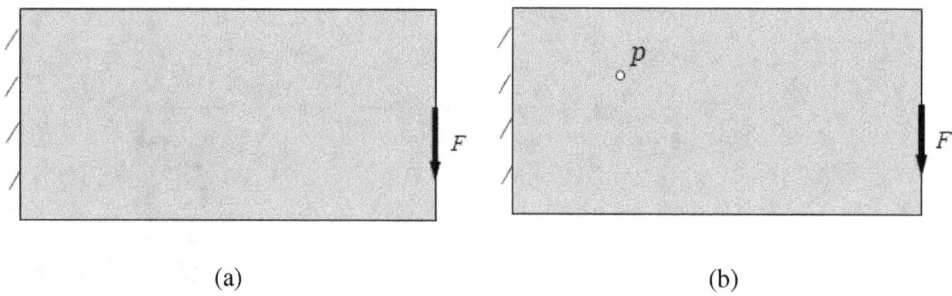

Figure 4.2: (a) A structural problem and (b) a topological change.

To compute the topological sensitivity field, consider the problem posed in Figure 4.3, where the radius of the hole is increased through a shape parameter τ.

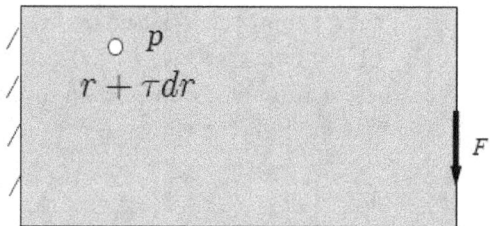

Figure 4.3: A hypothetical shape change.

One can now define the shape sensitivity of φ with respect to the shape parameter τ as:

$$\chi(r) \equiv \left. \frac{d\varphi(r)}{d\tau} \right|_{\tau=0} \qquad (4.23)$$

Well-known adjoint methods can be used to find closed-form expressions for shape sensitivity (for a fixed radius r). It has been established that topological sensitivity as defined in Equation 4.22 is related to shape sensitivity via:

$$\mathcal{T}(p) = \lim_{r \to 0} \frac{\chi(r)}{2\pi r} \qquad (4.24)$$

Consequently, a closed-form expression for the topological derivative can be derived. For example, in 2-D, it is given by:

$$\mathcal{T}(p) = \frac{4}{1+\nu}\sigma(\mathbf{u}) : \varepsilon(\boldsymbol{\lambda}) - \frac{1-3\nu}{1-\nu^2}tr(\sigma(\mathbf{u}))tr(\varepsilon(\boldsymbol{\lambda})) \qquad (4.25)$$

4.2 Level-set Methods

where ν is the Poisson ratio, $\sigma(u)$ is the stress field associated with the primary field u, and $\varepsilon(\lambda)$ is the strain field associated with an adjoint field λ that depends on the quantity of interest (the adjoint field λ is described below). Observe that the stresses and strains are evaluated at point p within the original domain before the hole is inserted. Similar expressions can be derived in 3-D.

4.2.2 Adjoint Field

The adjoint field λ in Equation 4.25 is defined as follows:

$$\mathbf{K}\boldsymbol{\lambda} = -\nabla_{\mathbf{u}}\varphi \tag{4.26}$$

For example, if the quantity of interest is the compliance, then:

$$\mathbf{K}\boldsymbol{\lambda} = -\nabla_{\mathbf{u}}(\mathbf{u}^T\mathbf{f}) = -\mathbf{f} \tag{4.27}$$

Thus, it follows that, for compliance:

$$\boldsymbol{\lambda} = -\mathbf{u} \tag{4.28}$$

The compliance topological sensitivity as computed via Equation 4.25 (and scaled for clarity) is illustrated in Figure 4.4.

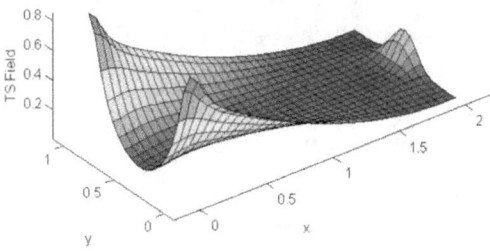

Figure 4.4: Topological sensitivity field $\mathcal{T}$ for compliance.

4.2.3 Pareto Method

To understand how the topological sensitivity field can be used for topology optimization, consider Figure 4.5a where the field is illustrated together with a cutting plane at an arbitrary height τ. Consider now a domain Ω^τ defined per:

$$\Omega^\tau \equiv \{p|\mathcal{T}(p) > \tau\} \tag{4.29}$$

In other words, the domain Ω^τ is the set of all points where the topological field exceeds the value τ; the induced domain Ω^τ is illustrated in Figure 4.5b.

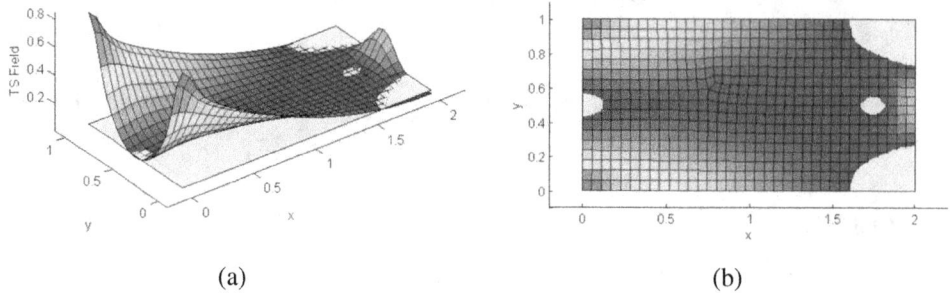

Figure 4.5: Compliance topological sensitivity field: (a) Cutting plane. (b) Induced domain Ω^τ.

This corresponds to a topology of reduced volume fraction such that the material contributing least to the stiffness of the structure is removed. The cutting-plane value τ can be chosen such that, say, 10% of the volume is removed. Of course, in practice, a more sophisticated algorithm is needed to ensure robustness and stability. This was carried out through the "Pareto" algorithm described below.

> Algorithm 4.2.1 **The Pareto algorithm**
> 1. The algorithm starts with $\Omega = D$, i.e., at a volume fraction of 1.0.
> 2. Next, a finite element analysis is executed on D and the topological sensitivity field is computed as discussed earlier.
> 3. If the desired volume fraction is reached, the iso-surface with the current cutting-plane value τ is extracted. For iso-surface extraction, we rely on the classic *marching-cubes algorithm*, where the topological sensitivity values at the nodes of the mesh are used to extract the iso-surface.
> 4. (Else) The current volume fraction is decremented by Δv; Δv is initialized to 0.05, and is controlled in an adaptive fashion (see step 8 below).
> 5. The parameter τ is computed such that $|\Omega^\tau|$ is equal to the current target volume fraction.
>
>> (R) This is a simple binary-search algorithm where maximum and minimum values of the topological sensitivity field serve as the limits of the binary search. When a topology Ω^τ is extracted using Equation 4.29, elements are classified as either being in or out; partial elements, i.e., pseudo-densities, are avoided since they lead to stiffness matrices with large condition numbers.
>
> 6. Once the desired value of τ has been computed, a finite element analysis is carried out on Ω^τ and the topological sensitivity field is recomputed.
> 7. If the τ value has converged (to within user-defined accuracy) the algorithm returns to Step 3. If the parameter has not yet converged, it returns to Step 5, after performing the check below to ensure that the optimization process is not diverging.
> 8. If a very large step size Δv is specified by the user, the above process may diverge. If this is detected (by diverging values of compliance), the value of Δv is reduced

4.2 Level-set Methods

by a factor of 2, prior to returning to Step 5.

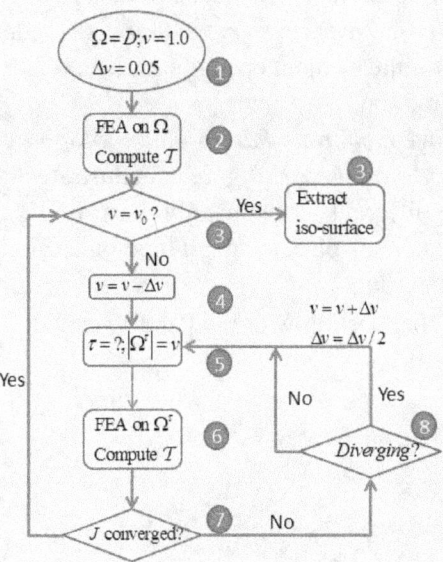

Figure 4.6: Pareto Algorithm.

4.2.4 Discussion

Pareto differs from SIMP in three ways:

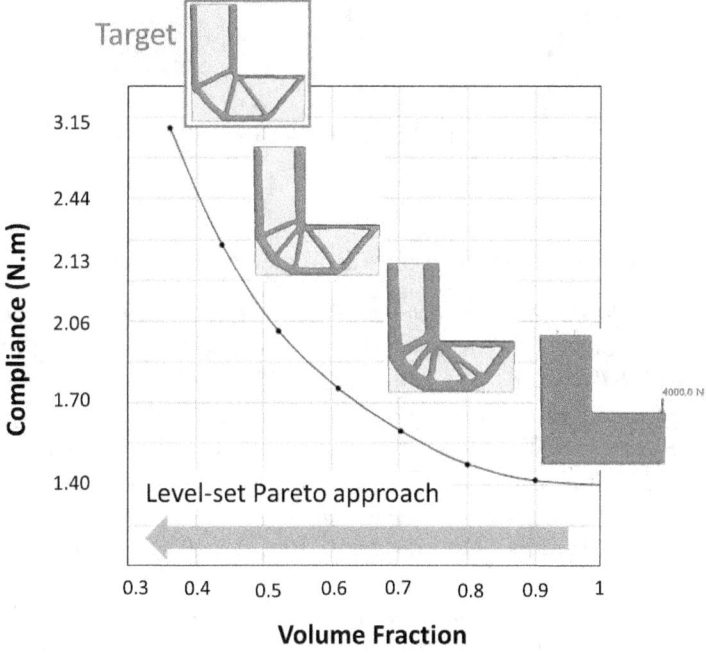

Figure 4.7: Topological sensitivity based Pareto tracing.

1. Pareto does not rely on pseudo-densities; consequently, the stiffness matrices are inherently better conditioned. This leads to faster convergence of iterative solvers, and, therefore, the *cost per finite element analysis* is reduced.
2. Typically, far fewer finite element operations are required, compared to SIMP, to reach a desired volume fraction.
3. Finally, one can find *numerous Pareto-optimal topologies* up to a desired volume fraction *with no additional cost*. Figure 4.7 illustrates the notion of Pareto-optimal topologies for two different problems. On the x-axis is the volume fraction, and on the y-axis is the relative displacement; all topologies along the curve are generated in a single optimization run.

Figure 4.8 illustrates the work-flow of the Pareto method.

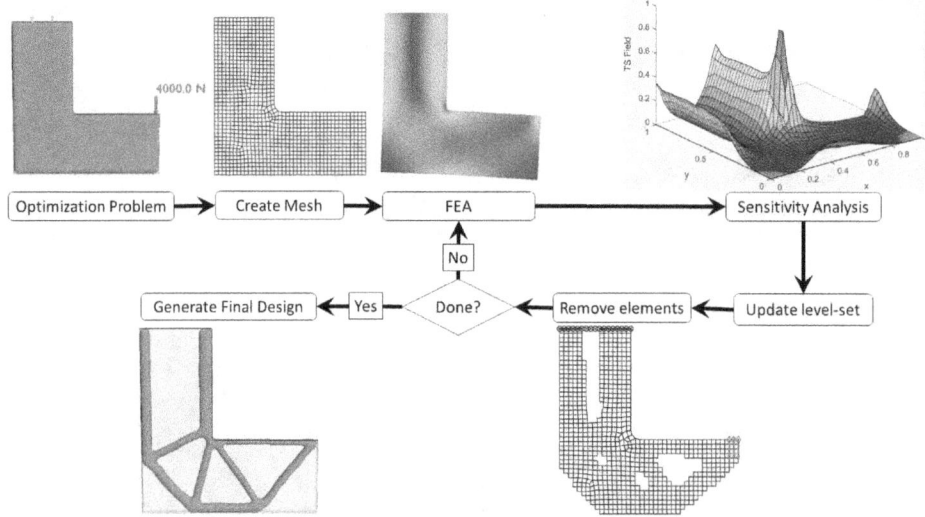

Figure 4.8: Level-set work-flow.

For additional description of the Pareto method, please see:

- *A 199-line Matlab code for Pareto-optimal tracing in topology optimization*, K. Suresh, Structural and Multidisciplinary Optimization 42, 665–679. doi:10.1007/s00158-010-0534-6, 2010,
- *Efficient generation of pareto-optimal topologies for compliance optimization*, I. Turevsky, K. Suresh, Int. J. Numer. Meth. Engng. 87, 1207–1228. doi:10.1002/nme.3165, 2011.
- *Efficient generation of large-scale pareto-optimal topologies*, K. Suresh, Struct Multidisc Optim 47, 49–61. doi:10.1007/s00158-012-0807-3, 2012.
- *Topology Optimization on the Cloud: A Confluence of Technologies*, K. Suresh, ASME 2015 International Design Engineering Technical Conferences and Computers and Information in Engineering Conference. American Society of Mechanical Engineers, p. V01AT02A041–V01AT02A041, 2015.

- *Stress-constrained topology optimization: a topological level-set approach*, K. Suresh, M. Takalloozadeh, Struct Multidisc Optim 48, 295–309. doi:10.1007/s00158-013-0899-4, 2013.
- *Multi-constrained topology optimization via the topological sensitivity*, K. Suresh, S. Deng Struct Multidisc Optim 51, 987–1001. doi:10.1007/s00158-014-1188-6, 2014.
- *A Pareto-Optimal Approach to Multimaterial Topology Optimization*, A.M. Mirzendehdel, K. Suresh, Journal of Mechanical Design 137, 101701–101701. doi:10.1115/1.4031088, 2015.
- *Support structure constrained topology optimization for additive manufacturing*, A.M. Mirzendehdel, K. Suresh, Computer-Aided Design 81, 1–13. doi:10.1016/j.cad.2016.08.006, 2016.

5. Getting Started with Pareto

There are several commercial topology optimization software available today which are capable of solving complex 3D problems. Most of these software rely on SIMP. In this text, we will use the Pareto software that relies on the Pareto technology developed at the University of Wisconsin-Madison. This chapter provides a short introduction to Pareto. Pareto has various instances, designed for different platforms, including ParetoWin, ParetoLinux, ParetoMac, ParetoCloud, and ParetoWorks (an add-in to SolidWorks). Their interface and capabilities are almost identical. For the remainder of this text, we will largely use ParetoWin.[1]

Table 5.1: Pareto Instances.

Pareto Instance	Platform	Type	Interface	Requirements	Limitations
Pareto	Windows MacOS Linux	Standalone	Integrated	-	None
ParetoCloud	Any	Cloud	Browser	Browser	Limited server time
ParetoWorks	Windows	Add-In	SolidWorks	SolidWorks	None

---

[1] Pareto is currently being commercialized through SciArt (`www.sciartsoft.com`). You can request for a trial license for any of the Pareto instances through SciArt.

5.1 Pareto

Topology optimization is computationally intensive, especially for multi-constrained, multi-load 3D problems. Pareto alleviates this issue by exploiting assembly-free FEA, implemented on multi-core platforms. It is therefore recommended that you use a computer with a multi-core CPU, with the minimum specifications of:
- i7 quad-core
- At least 8 GB RAM
- Preferably equipped with a NVidia or AMD Graphics Card

5.1.1 ParetoWin

To install ParetoWin, you can follow the steps described below:

1. Obtain an installer for ParetoWin by contacting support@sciartsoft.com.

Figure 5.1: ParetoWin Installer.

2. Proceed to install ParetoWin (Visual Studio Redistributable will also be installed).

Figure 5.2: ParetoWin Installer.

3. ParetoWin will be installed as a standard Windows Application and a *ParetoExamples*

5.1 Pareto

folder will also be created on your desktop.

Figure 5.3: (a) ParetoWin icon, (b) ParetoExamples folder.

4. If you do not have access to a permanent license, you will see the following message; press OK.

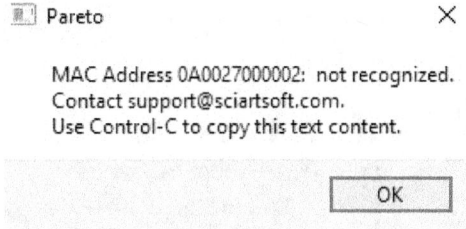

Figure 5.4: No permanent access warning.

5. Press OK to use the temporary license (until expiry).

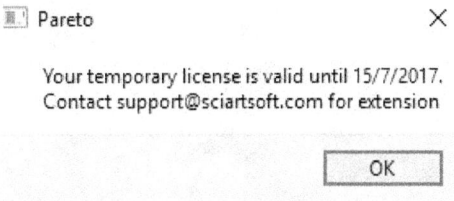

Figure 5.5: Trial license.

6. Once the installation is complete, you can run ParetoWin, whose interface is illustrated in Figure 5.6.

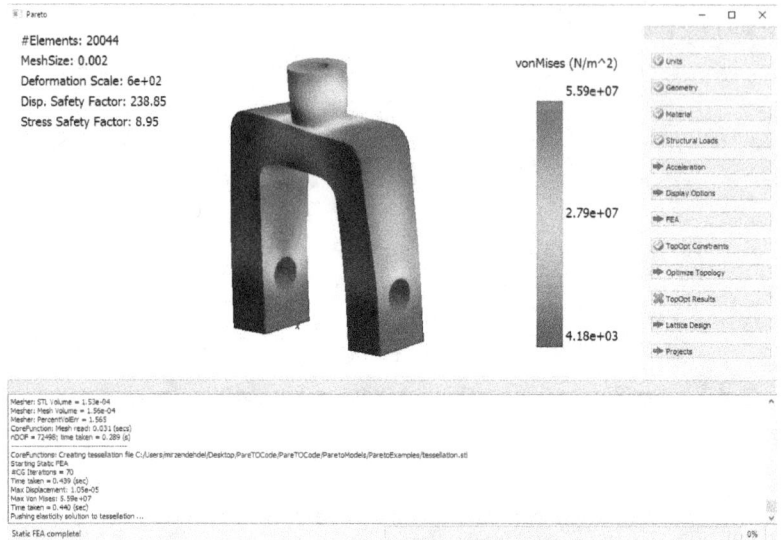

Figure 5.6: ParetoWin interface.

Illustrative examples are provided later in this chapter.

5.1.2 ParetoMac

You can obtain an installer for ParetoMac by contacting support@sciartsoft.com. ParetoMac installation is similar to ParetoWin, except that you will start with a .dmg file. Once the installation is complete, you can run ParetoMac, whose interface is illustrated in Figure 5.7.

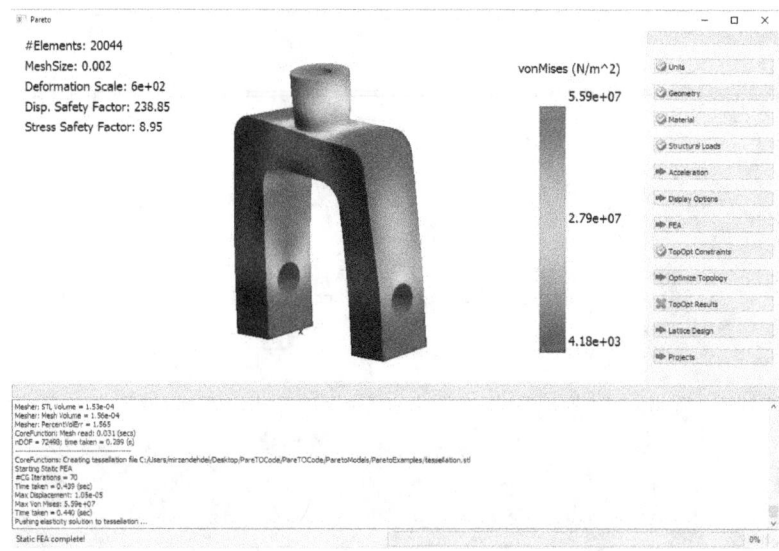

Figure 5.7: ParetoMac interface.

5.1.3 ParetoWorks

If you have a SOLIDWORKS™ license, you can choose to use ParetoWorks, instead of ParetoWin. ParetoWorks is an Add-in to SOLIDWORKS, and its interface is illustrated in Figure 5.8. Contact support@sciartsoft.com for a trial license of ParetoWorks.

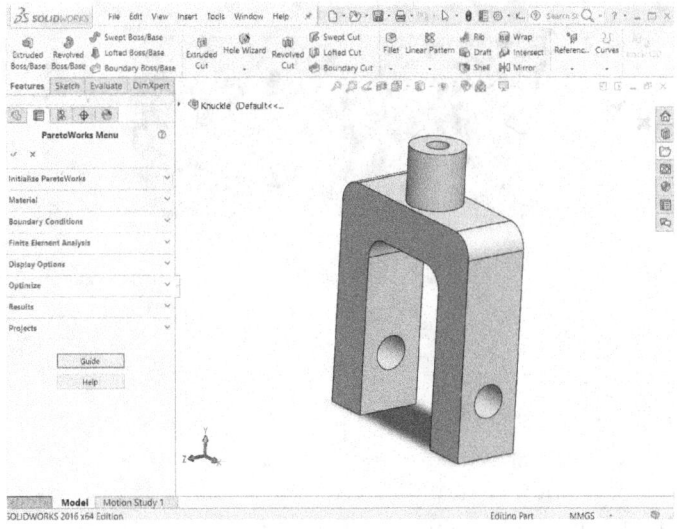

Figure 5.8: ParetoWorks interface.

5.1.4 ParetoCloud

ParetoCloud is a web-based implementation of Pareto, and you only need a browser (such as Internet Explorer, Chrome, Firefox, or Safari) to access the software; there is nothing to download. You can access ParetoCloud through www.cloudtopopt.com; the interface is illustrated in Figure 5.9. Video tutorials are available through the software website.

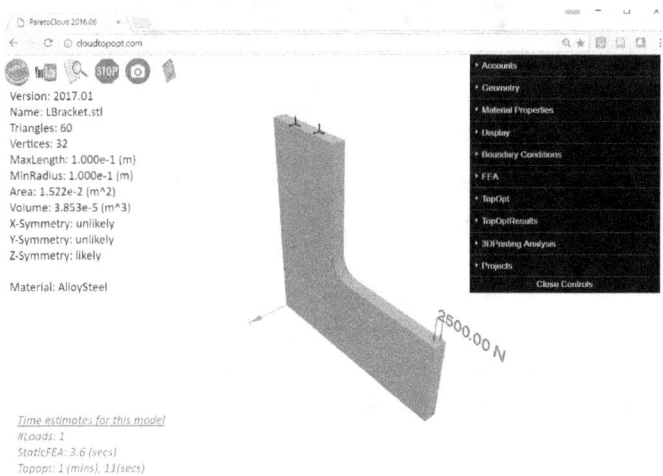

Figure 5.9: ParetoCloud interface.

To request a new account:

1. Open a browser of your choice and go to www.cloudtopopt.com
2. Under *Accounts*
    - Go to *Request Account*
    - Enter your *Email* address and press *Send Request*.
    - A username and password will be sent to your email address.

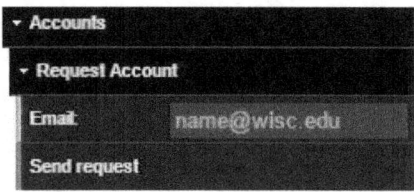

Figure 5.10: ParetoCloud Sign-up.

To sign in to your account:
1. Under *Accounts*, go to *Sign In*
2. Enter your *UserName* and *Password*

> If you have just signed-up, your balance should include 100 seconds of computation (bottom left of your screen). If you have run out of computation time, you can *Replenish Balance* to 100 seconds but it would take **1 hour** before you can use it.

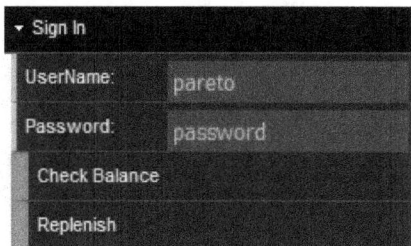

Figure 5.11: ParetoCloud Sign-in.

5.2 Examples

We will now illustrate the use of Pareto through a series of examples. For simplicity, we will use ParetoWin; however, these examples carry over to other instances of Pareto.

The first step in topology optimization is to pose a valid finite element problem. Specifically, within the context of structural analysis, one must define the geometry, select a material, apply restraints and forces in a meaningful way.

▪ **Example 5.1 Posing a valid structural problem**

1. In ParetoWin, under the *Units* menu, select MKS and apply.

5.2 Examples

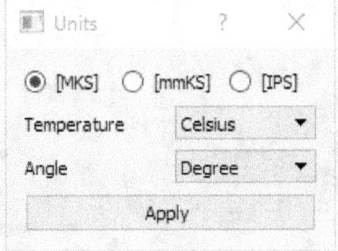

Figure 5.12: Apply SI units.

2. Under the *Geometry* menu, click on *Load STL*. Then, from the *ParetoExamples* folder, select the *ThreeHoleBracket.stl* file; the model will be loaded as in Figure 5.13.

(a)   (b)

Figure 5.13: (a) Load STL, (b) Three-hole bracket model

(R) Currently, ParetoWin can read geometry only in .stl format.

3. Observe (in the left-hand top corner) of the graphics window that no material has been applied; see Figure 5.14.

> Model: ThreeHoleBracket
> Length: 0.10 (meter)
> Part 0: NoMaterial
> #LoadSets: 0

Figure 5.14: No material has been applied initially.

4. Under the *Material* menu, select the default material of *Alloy Steel* and Apply as in Figure 5.15.

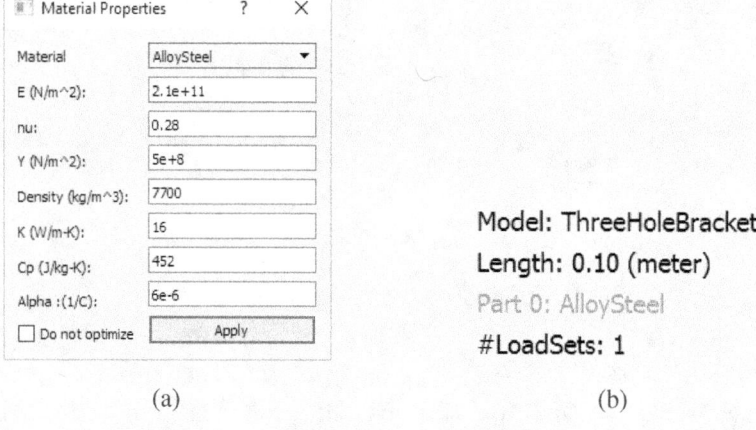

(a)   (b)

Figure 5.15: Applying Alloy Steel.

(R)   Material properties can be modified before applying. For example, the yield strength is set to 500 MPa; this can be lowered to an *allowable stress limit* (with a built-in safety factor). Material application will be reflected on the graphics window as in Figure 5.15.

5. Next, we will apply restraints on the model:
   - Under *Structural Loads*, set *Selection* to *Coarse Cylinder*.
   - Select the top left cylindrical hole using the Alt-key and left button on your mouse (see Figure 5.16a). To unselect a previously selected face, use the Alt-key and right button.
   - Then, select the bottom left hole as in Figure 5.16b.

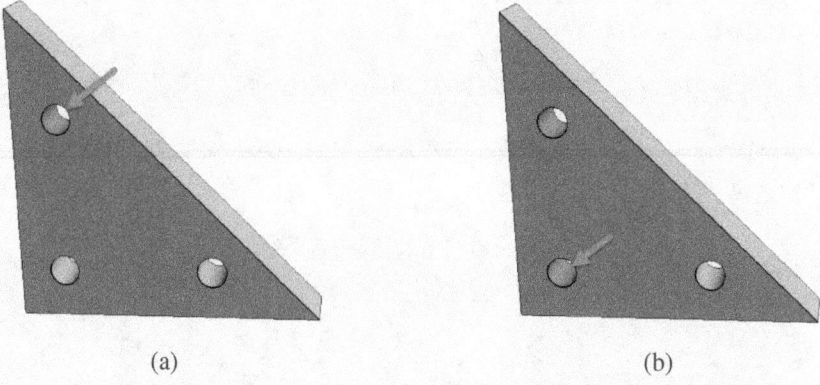

(a)   (b)

Figure 5.16: (a) First face selection, (b) second face selection.

- Next, set the *Load Type* to *XYZFixed* as in Figure 5.17a, and Apply; this will restrain the 2 faces in all three directions; see Figure 5.17b.

5.2 Examples

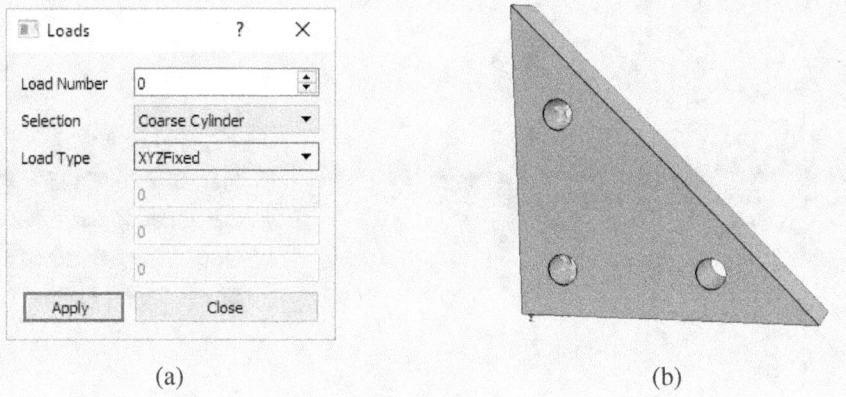

(a)   (b)

Figure 5.17: (a) restraint selection, and (b) restraint conditions.

6. Next, we will apply a force on the third hole. Select the face as illustrated in Figure 5.18a:
   - Change the *Load Type* to Force.
   - Set the Y component to -60,000 N. You can change the display to transparent to see the force and restraints.

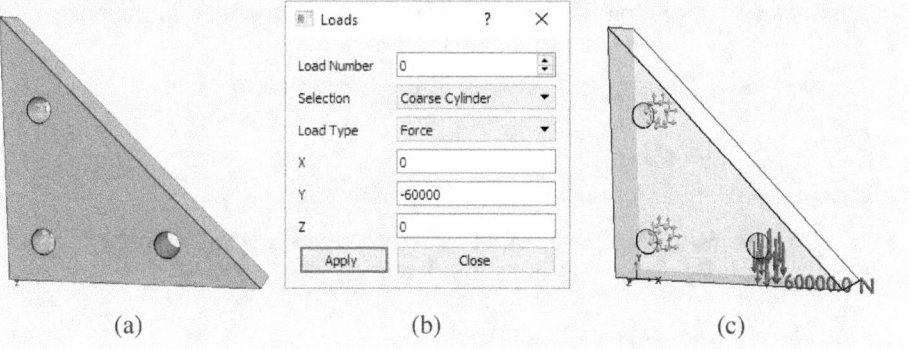

(a)   (b)   (c)

Figure 5.18: (a) Select the third face, (b) apply a load, and (c) resulting condition.

Once the structural problem is posed, we can carry out an FEA.

■ Example 5.2   FEA of the three-hole bracket

1. Under *FEA* menu, set *Quality* to medium mesh quality.
2. Select the *Z Symmetry* option; the algorithm will attempt to maintain symmetry along z-direction.
3. Carry out static FEA.

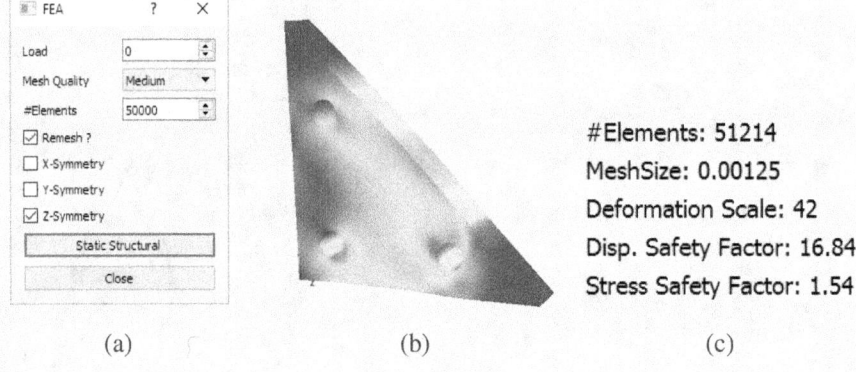

| (a) | (b) | (c) |

Figure 5.19: (a) Medium mesh quality, (b) stress results, and (c) safety factors.

Based on the yield strength for alloy steel, we observe that the stress safety factor is 1.54; see Figure 5.19c (disregard the large displacement safety factor).

We will now carry out topology optimization for the above structural problem.

■ Example 5.3 Topology optimization of the three-hole bracket

1. Under *TopOpt Constraints* menu:
2. Set relative minimum feature size *RelMinFeatSize* to 2.

   RelMinFeatSize corresponds to smoothness of sensitivity field. Higher values would eliminate thin features.

3. Set *Stress Safety Factor* to 1.00.
4. Set *Displacement Limit* to 0.002 (m).
5. Select the *Keep Fixed Faces* option; this means that the material surrounding the two restrained holes will be retained during optimization.
6. Apply.

5.2 Examples

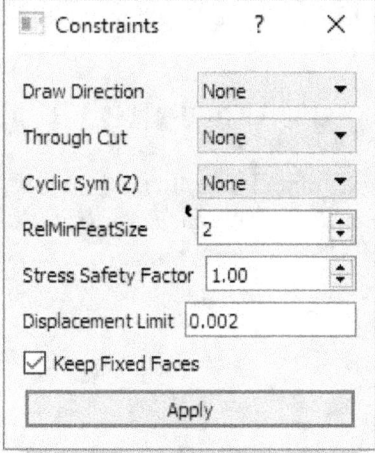

Figure 5.20: Topology optimization constraints.

7. Under *Optimize Topology* menu (Figure 5.21a), set the desired volume fraction to 0.5. The objective is set to the default of objective of minimizing compliance.
8. Optimize.

Upon optimization, the final model is illustrated in Figure 5.21b, and the final safety factors are illustrated in Figure 5.21c. Observe that we have not reached the yield strength, i.e., the safety factor is greater than 1. This is due to the fact that we set the desired (final) volume fraction to be 0.5.

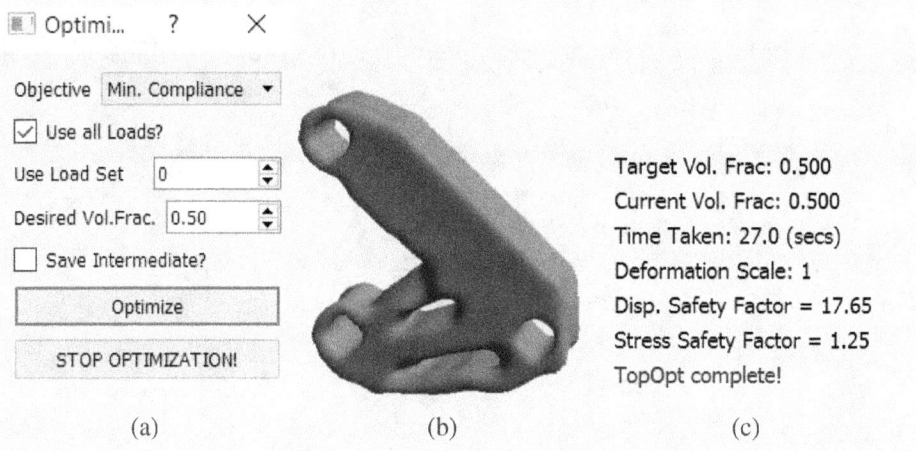

Figure 5.21: (a) Optimization parameters, (b) optimized model, (c) safety factors.

⓿ The optimization algorithm in Pareto is regularly updated; so the final design you obtain may differ from the one in Figure 5.21. The non-uniqueness of optimized designs in TO is discussed in a later chapter.

You can now save the work including the material, restraints and loads as described below.

■ **Example 5.4 Saving the work as a project file**

1. Under *Projects* menu (Figure 5.22), click on Save current project
2. When prompted, save the project as *ThreeHoleBracketProject.prj* in the same directory as *ThreeHoleBracket.stl*. You can use any other name for the project, if you do not wish to overwrite a file that already exists with that name.

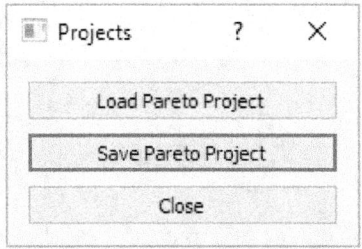

Figure 5.22: Saving a project file.

(R) The project file must be saved in the same directory as the .stl file.

In Pareto, one can also recover several intermediate topologies at higher volume fractions.

■ **Example 5.5 Recovering intermediate Pareto-optimal designs**

1. Under the *Optimization Results* menu (Figure 5.23a, change the *Topology@Vol* to 0.6 (Figure 5.23b). The corresponding topology will be displayed.

5.2 Examples

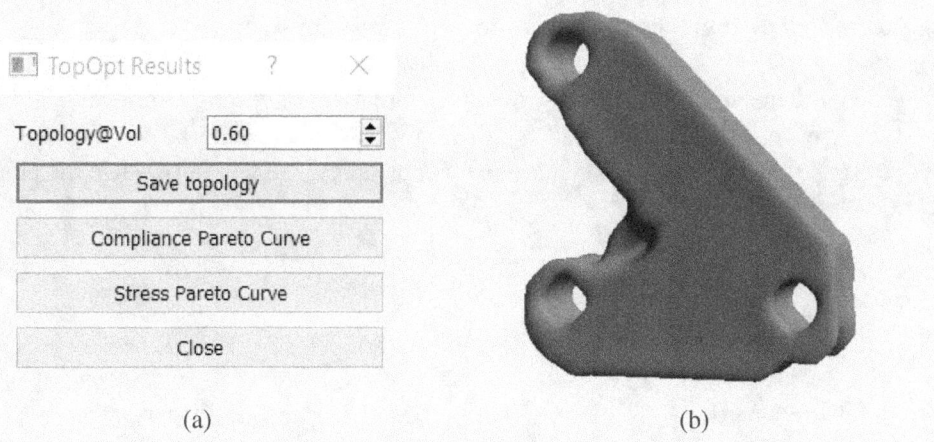

(a)   (b)

Figure 5.23: Recovering intermediate topologies, (a) Results menu, and (b) optimized design at 0.60 volume fraction.

2. These intermediate topologies are Pareto optimal in that they lie on the Pareto curve, hence the name "Pareto" for the algorithm. Select the compliance curve button in Figure 5.23a. Figure 5.24 illustrates the progression of the optimization process up to a volume fraction of 0.5. Observe that the optimization begins with a volume fraction of 1.0.

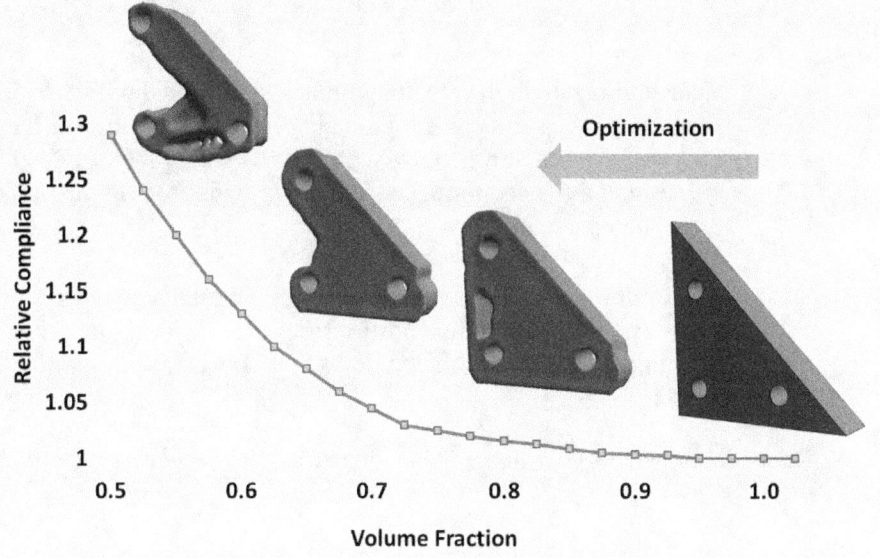

Figure 5.24: Pareto curve for three-hole bracket optimization.

We will now repeat the topology optimization, but with a stress constraint.

■ Example 5.6 Topology optimization with stress constraint

1. Under the *Optimize Topology* menu (Figure 5.25a), set the desired volume fraction is set to 0.01, i.e., we will try to remove as much material as possible, while satisfying the stress and displacement constraints. We will not change the objective.
2. Optimize.

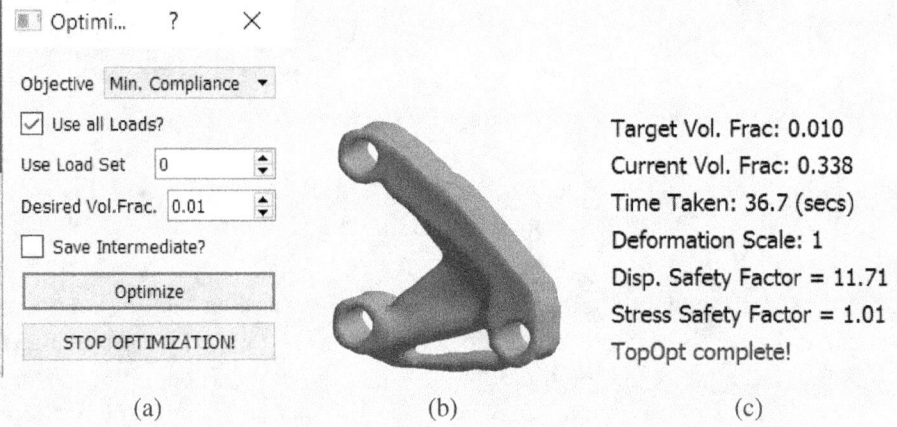

(a)   (b)   (c)

Figure 5.25: (a) Optimization parameters, (b) optimized model of 0.34 volume fraction, (c) safety factors.

- (R) Upon optimization, the resulting model is illustrated in Figure 5.25b, and the safety factors are illustrated in Figure 5.25c. Observe that the final volume fraction is 0.34, with a stress safety factor of 1.0, i.e., the design has been optimized to the lowest possible volume fraction, given the constraints.

- (R) The optimization algorithm in Pareto is regularly updated; so the final design you obtain may differ from the one in Figure 5.25. The non-uniqueness of optimized designs in TO is discussed in a later chapter.

3. Under *Display Options*, change the display mode to *Final Stress* (Figure 5.26).

5.2 Examples

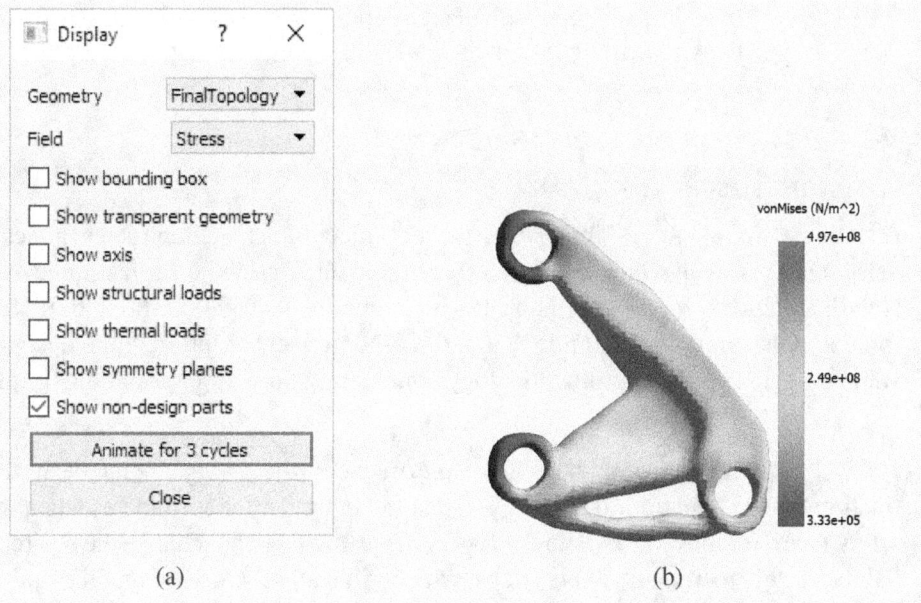

Figure 5.26: (a) Changing the display mode to *FinalStress*, and (b) the stress distribution.

4. The final design with a stress safety factor of 1.0 might be of concern to the reader. However, note that one can always recover designs with higher safety factors as illustrated earlier. Alternately, one can simply set a higher safety factor prior to optimization.
5. Figure 5.27 illustrates a few Pareto optimal designs, and their safety factors.

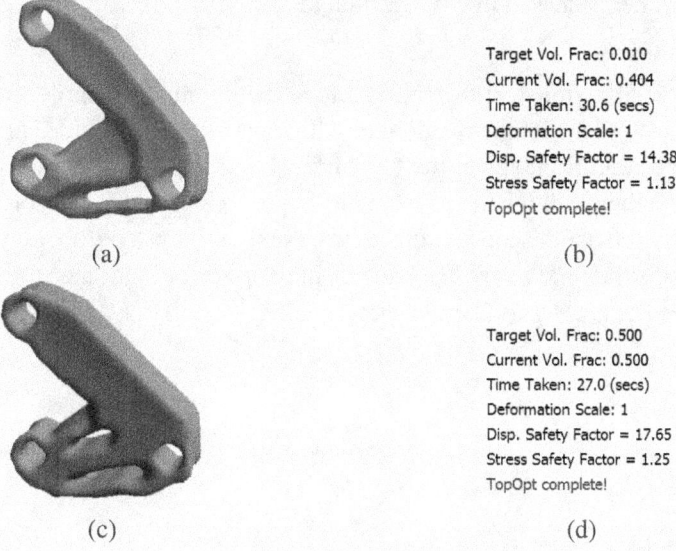

Figure 5.27: Exploring the Pareto-optimal designs.

(R) In addition to recovering Pareto-optimal designs, one can also modify the yield strength to include a built-in safety factor.

5.3 Exercises

Exercise 5.1 Upload the ThreeHoleBracket project file created in the examples section. Select the surface on which a force has been applied; replace the vertical force of -60,000 N with a diagonal force with the following components (60,000,-60,000,0) N, i.e., the force now has both an x-component and a y-component. Optimize the design for 0.5 volume fraction. Is the optimal topology consistent with your expectation? Explain.

Exercise 5.2 Upload again the ThreeHoleBracket project. Change the desired volume fraction to 0.01, and optimize. Observe that the optimization terminates when the stress safety factor reaches 1. Next, under TopOpt Constraints menu, change the desired "Stress Safety Factor" to 1.5, and apply. Repeat the optimization. Observe that the optimization terminates when the stress safety factor reaches 1.5. In a future chapter we will discuss the impact of such performance constraints.

Exercise 5.3 Upload again the ThreeHoleBracket project. Under TopOpt Constraints menu, change the "ThroughCut" option to z-direction, and apply. In other words, we the design must exhibit constant thickness along z-direction. Repeat the optimization. Observe that the resulting topology indeed exhibits a constant thickness, and can be fabricated using a through-cut operation. The impact of such manufacturing and design constraints will be discussed in a future chapter.

Exercise 5.4 Set the units set to mm, and load the geometry CantileverBeam.stl from ParetoExamples folder. Set the material to AlloySteel. Select the side surface, and apply fixed boundary condition; select the top surface, and apply a pressure of 10 units; see Figure 5.28. Set the mesh quality to medium; impose z symmetry and solve the structural problem. Apply the default topopt constraints. Next optimize the topology to 0.5 volume fraction. What is the final stress? Use the display options to plot the stress over the final topology. Save the project as CantileverBeamExercise.

5.3 Exercises

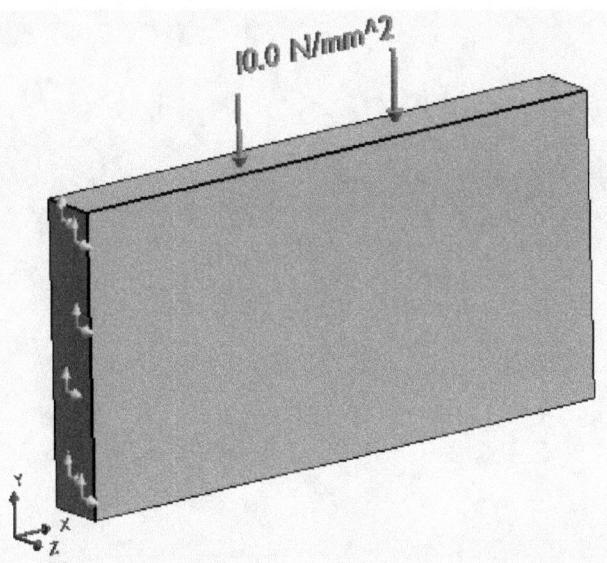

Figure 5.28: Structural problem posed over a cantilever beam.

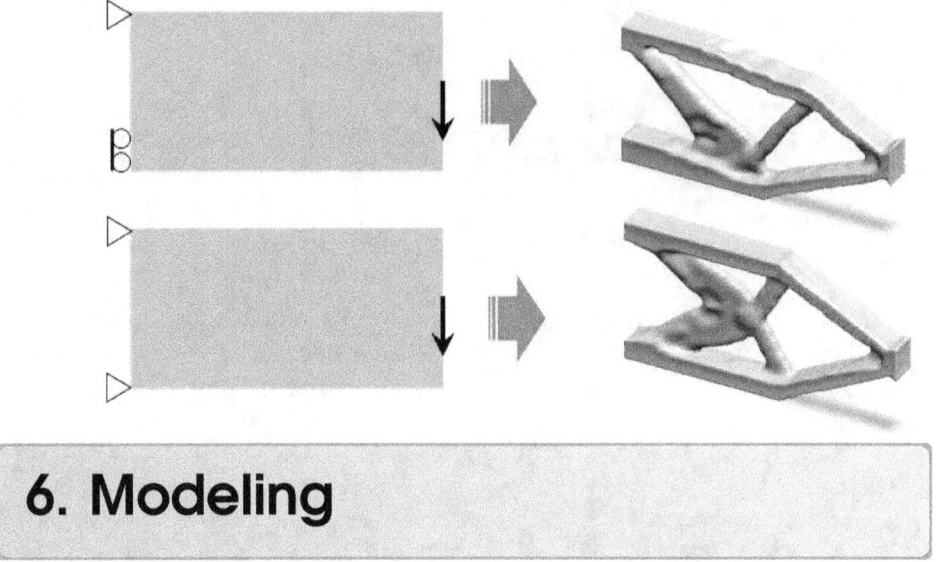

6. Modeling

The first step in topology optimization is *modeling*. Modeling is the process of converting a loosely-worded problem into a valid and unambiguous topology optimization problem, i.e., one must define the design space, material, boundary conditions, manufacturing constraints, etc. Modeling will invariably involve certain assumptions. Needless to say, an incorrect assumption, or an oversight, will lead to an incorrect topology. To illustrate, we consider a few examples in this chapter.

6.1 Table Design

First, we will study the impact of restraints on topology optimization. As an example, we will consider the design of a table under two different types of restraints.

■ **Example 6.1 Table Design**

1. Under the *Units* menu, select in-lb-sec units.
   - ⓡ Note that one must be aware of the units that the .stl file was saved under (from your CAD system), and the same units must be used here. In this case, the table model was saved in inches.
2. Under *Geometry*, select *Load STL*
3. From the *ParetoExamples* folder, select the *Table.stl* file; the model will be loaded as in Figure 6.1.
4. Under the *Material* menu, select *Al 1060 (Aluminum)* material, and apply.

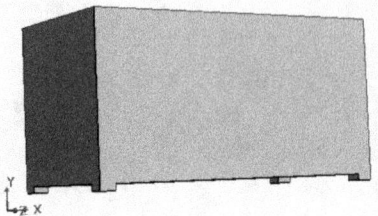

Figure 6.1: Table model.

5. Next, we will apply the restraints. Select one of the supports and apply a fixed restraint as in Figure 6.2a. Recall that to select a surface, use the Alt-key and left button, and to unselect, use the Alt-key and right button. The resulting display is shown in Figure 6.2b.

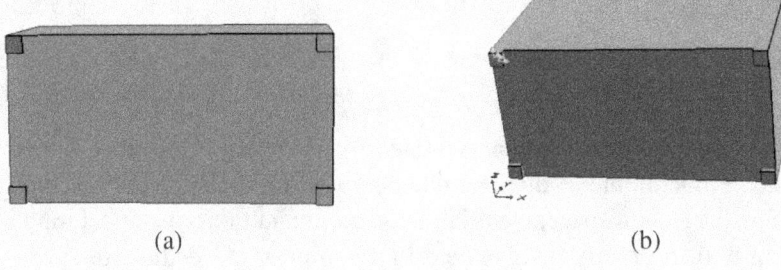

(a)   (b)

Figure 6.2: (a) Applying a boundary condition on one support (c) Fixed boundary condition on one support.

Next select the remaining three supports, and apply a sliding restraint as in Figure 6.3.

Ⓡ   The sliding condition allows the table to slide along the floor, without over-constraining it.

6. To see the final set of restraints, turn on transparent mode (under *Display* menu), as in Figure 6.3(b).

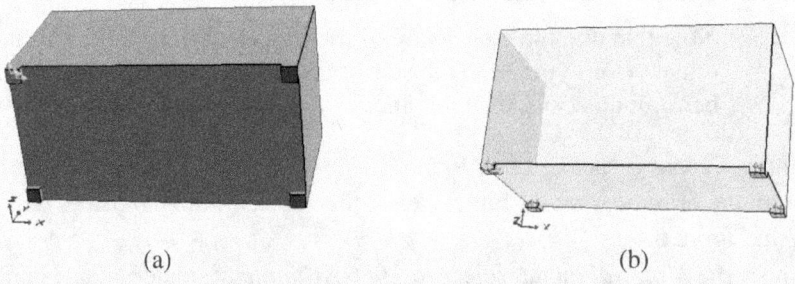

(a)   (b)

Figure 6.3: (a)Sliding boundary condition on the three remaining supports, (b) all restraints applied on the table.

6.1 Table Design

7. Select the top face, and apply a normal force of 1000 lbs as in Figure 6.4.

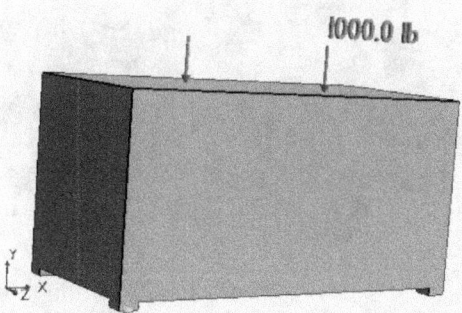

Figure 6.4: A normal force of 1000 lb.

8. Set mesh quality to medium; set X and Z symmetries, and solve the static problem.
9. The deflection plot is illustrated in Figure 6.5. Use the animation button to animate the displacement; observe how the three supports are moving relative to the fixed support.

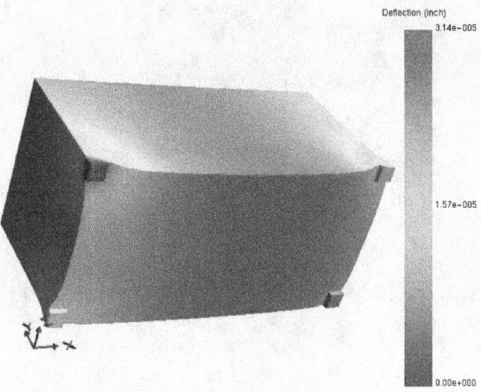

Figure 6.5: Deflection plot.

10. We will now carry out optimization with additional displacement constraints. Set the optimization parameters as follows:
    - Maximum displacement of 0.79 (in).
    - Keep fixed faces.
11. You may want to save the project before optimizing.
12. Now carry out the optimization with a target volume of 0.2. The optimized design is illustrated in Figure 6.6.

Figure 6.6: Optimized design with sliding constraints.

13. The final displacement plot is illustrated in Figure 6.7. Once again, observe how the supports are moving relative to the fixed support.

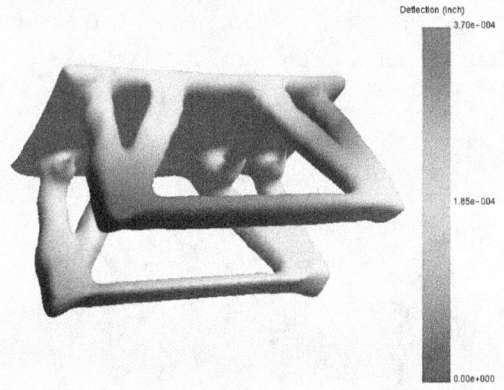

Figure 6.7: Displacement plot of optimized design.

14. Under *Display Options* menu, you can also animate the deformation using the Animate button.

We will now change the sliding restraints to fixed restraints. This is equivalent to assuming that either: (a) the table legs are firmly bolted to the ground, or (2) the friction force is sufficiently high that the legs do not slide relative to one another.

■ **Example 6.2 Table design restraints revisited**

1. Under Display, switch to initial design with field set to none. Select to show the structural loads.
2. Select the three faces that were previously set to sliding condition as in Figure 6.8(a); set the boundary condition to fixed.
3. The resulting boundary conditions are illustrated in Figure 6.8(b).

6.1 Table Design

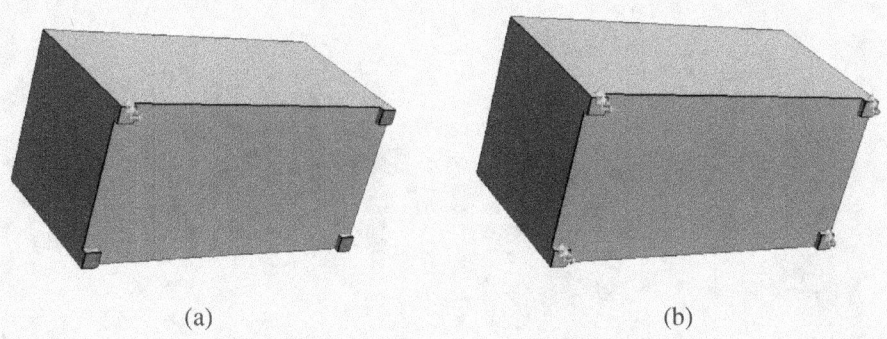

(a) (b)

Figure 6.8: (a) Restraining the three supports with fixed constraints, (b) all supports are now fixed.

4. The deformation is illustrated in Figure 6.9. One can animate the FEA results to observe that all supports remain fixed.

Figure 6.9: Deformation with fixed supports.

5. Next optimize with the same set of parameters as before; the resulting topology is illustrated in Figure 6.10(a).
6. The deformation plot is illustrated in Figure 6.10. Comparing the final topology against the one in Figure 6.6, observe that the additional horizontal connectors are not needed now since the legs do not move relative to one another. Further, the final displacement field is symmetric along X and Y.

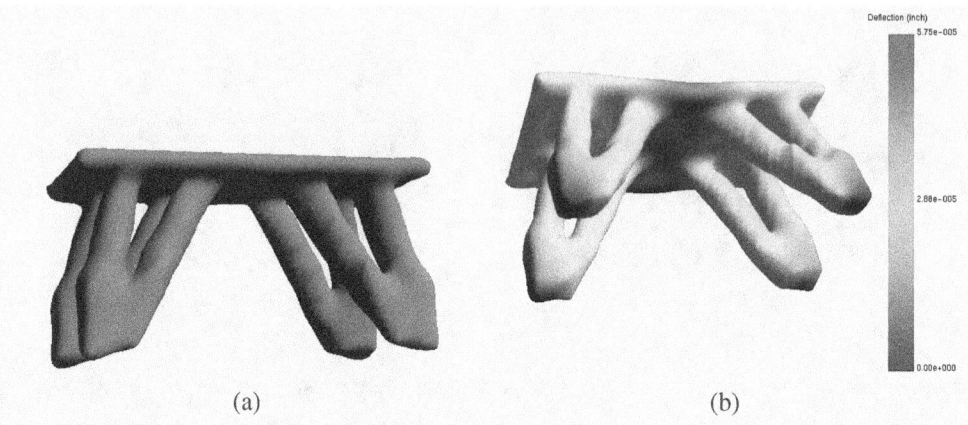

(a)   (b)

Figure 6.10: (a) Optimized design with fixed constraints, (b) Displacement plot with fixed constraints.

6.2 Knuckle Design

As a second example, we will consider the design of a knuckle under different restraints.

■ Example 6.3 Knuckle Design

1. Under the *Units* menu, select m-N-sec units.

   ® Note that one must be aware of the units that the .stl file was saved under (from your CAD system), and the same units must be used here. In this case, the knuckle model was saved in meters.

2. Under *Geometry*, select *Load STL*
3. From the *ParetoExamples* folder, select the *Knuckle.stl* file; the model will be loaded as in Figure 6.11.

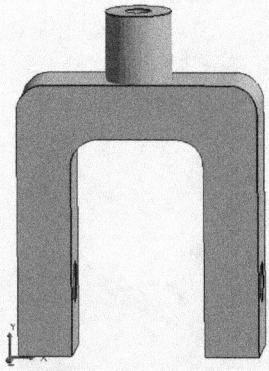

Figure 6.11: Knuckle model.

4. Under the *Material* menu, select *Alloy Steel*, and apply.
5. Next, select one of the bottom cylindrical features, and apply a fixed restraint. The resulting plot is illustrated in Figure 6.12a. Next select the other cylindrical hole, and apply a sliding restraint as in Figure 6.12b.

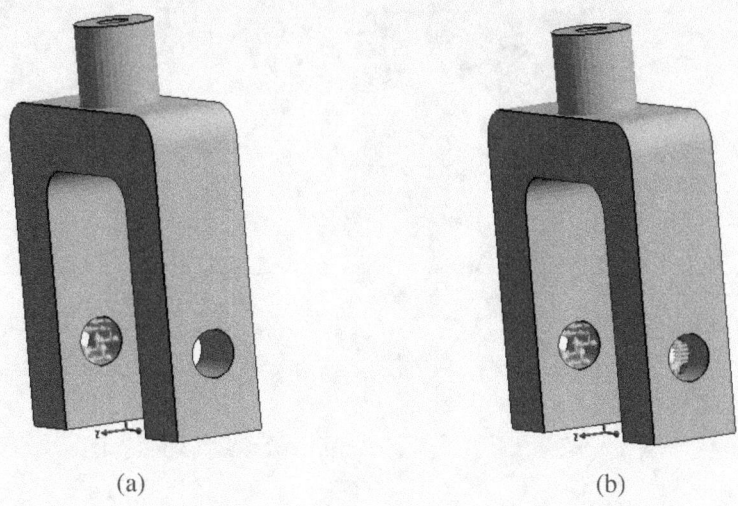

Figure 6.12: (a) Fixed boundary condition on one feature. (b) Sliding boundary condition on the other.

6. Select the top cylindrical feature, and apply a vertical force (y-direction) of 5000 N as in Figure 6.13.

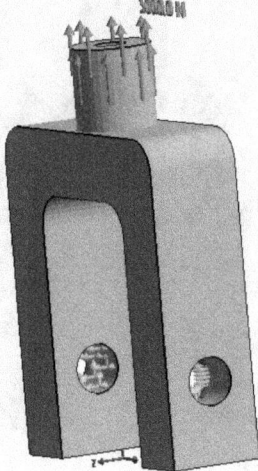

Figure 6.13: A vertical force of 5000 N.

7. Set mesh quality to coarse, and solve the static problem. The stress plot is illustrated in Figure 6.14. If you animate the result, you will notice how one of the supports moves relative to the other.

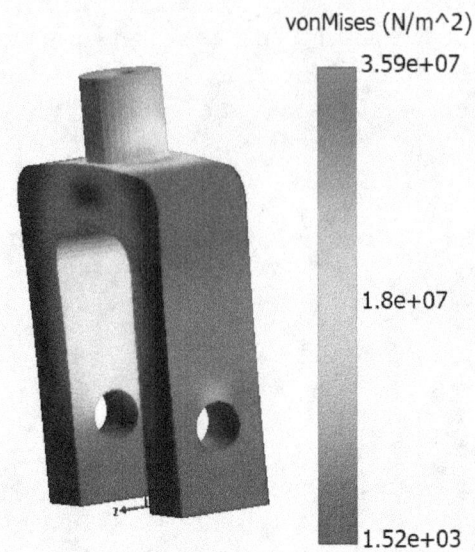

Figure 6.14: Stress plot for the knuckle design.

8. Set the optimization parameters as follows:
    - Choose to keep the fixed faces.
9. Now carry out optimization for a target volume fraction of 0.5. The optimized design is illustrated in Figure 6.15.

Figure 6.15: Optimized design with sliding constraint.

10. Under *Display Options* menu, you can also animate the deformation using the Animate option.

We will now change the sliding restraint to a fixed restraint. Once again, the right choice depends on how the two holes are actually restrained.

6.2 Knuckle Design

■ **Example 6.4 Knuckle design restraints revisited**

1. Under Display, switch to initial design with field set to none. Select "show structural loads".
2. Select the face that was previously set to sliding condition, and change the boundary condition to fixed.
3. The resulting boundary condition is illustrated in Figure 6.16.

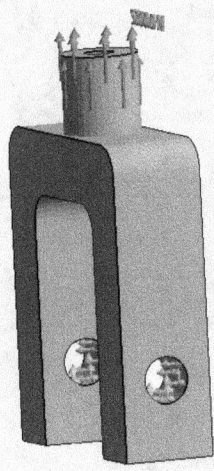

Figure 6.16: Knuckle model.

4. You can choose to carry out an FEA, and animate the results.
5. Next optimize with the same set of parameters as before; the resulting topology is illustrated in Figure 6.17. Once again, observe the significant difference in topology.

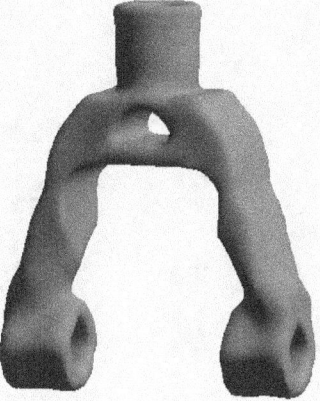

Figure 6.17: Optimized design with fixed constraints.

The main conclusion to draw from the above two examples is that a fully restrained boundary condition will lead to a different topology compared to a sliding restraint. One should therefore apply boundary conditions carefully.

6.3 Spindle Mount Design

We will now consider modeling and optimizing the topology of a spindle mount. A typical spindle mount assembly is illustrated in Figure 6.18.

Figure 6.18: A typical spindle assembly.

Figure 6.19a illustrates a CAD mock up of the spindle mount assembly, while Figure 6.19b illustrates typical forces and torques acting on the spindle. We will assume that the spindle mount is restrained as illustrated. Further, we will isolate the spindle mount, and transfer the forces and torques from the spindle to the spindle mount. In a later chapter, we will model the entire assembly.

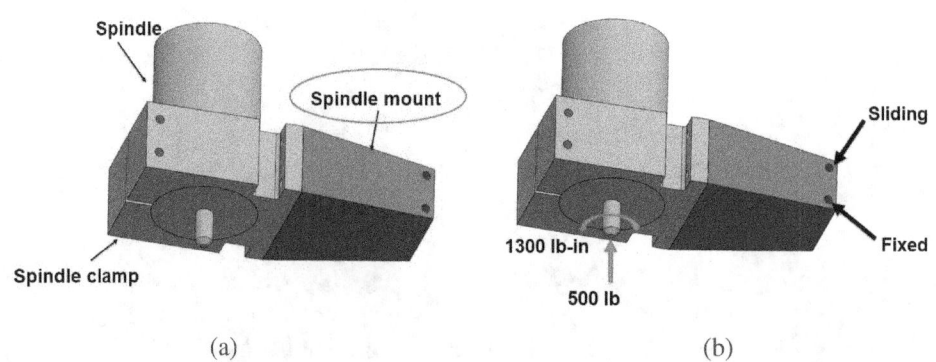

Figure 6.19: (a) Spindle mount, and (b) Forces, torques and restraints.

6.3 Spindle Mount Design

■ Example 6.5 Spindle Mount Design

1. Under the *Units* menu, select in-lb-sec units.
2. Under *Geometry*, select *Load STL*
3. From the *ParetoExamples* folder, select the *SpindleMount.stl* file; the model will be loaded as in Figure 6.20.

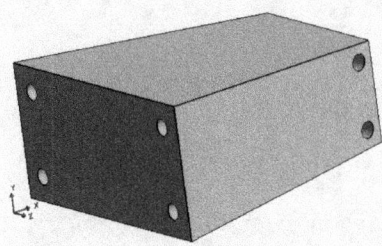

Figure 6.20: Spindle mount model.

4. Under the *Material* menu, select *Al1060*, change the yield strength to 30 ksi and apply. Note that this will override the default yield strength.
5. Next, apply the restraints as in Figure 6.19b, i.e., one of the through holes is fixed, while the other is sliding. The resulting plot is illustrated in Figure 6.21.

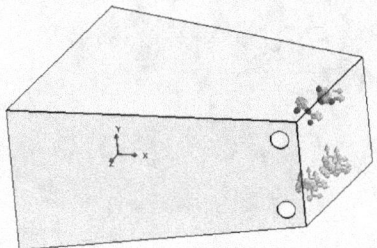

Figure 6.21: Spindle mount model.

6. We will disregard the 1300 lb-in torque and the moment caused by the 500 lb force. Instead we will simply distribute the 500 lb force among the four mounting holes. In the next example, we will correct this modeling error.
7. Select one of the four mounting holes, and apply a vertical force (y-direction) of 125 lb as in Figure 6.22a.
8. Repeat this individually for each of the remaining three holes. Note that applying individual forces will make it easier to edit them later on; see Figure 6.22b.

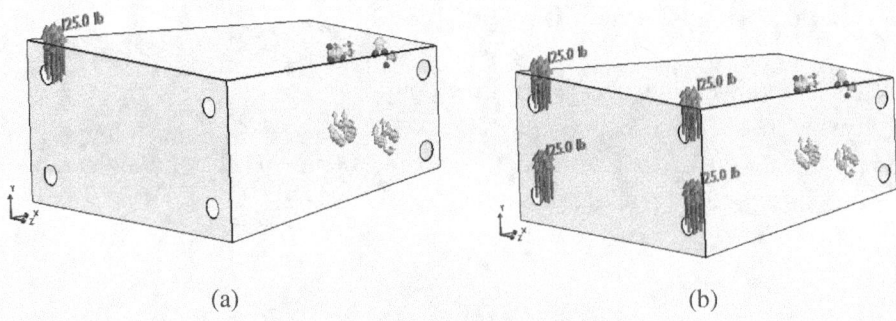

(a)   (b)

Figure 6.22: The transfer of the 200(a) One of the , and (b) Spindle mount dimensions (inches).

9. Set mesh quality to medium, enforce z symmetry, and solve the static problem. The stress plot is illustrated in Figure 6.23.

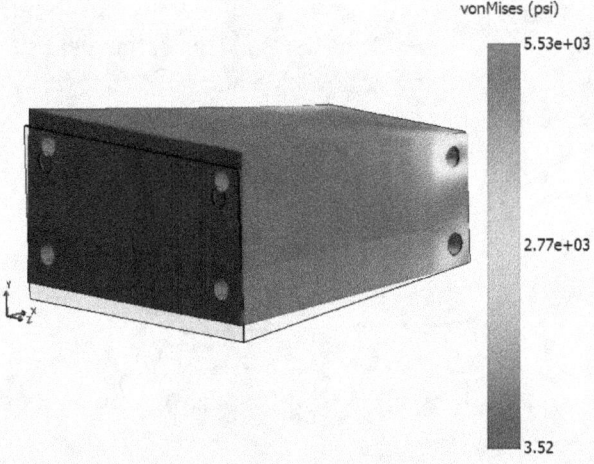

Figure 6.23: Stress plot for the spindle mount design.

10. Set the optimization parameters as follows:
    - Choose to keep the fixed faces.
11. Now carry out optimization for a target volume fraction of 0.3. The optimized design of 0.3 volume fraction is illustrated in Figure 6.24. The stress safety factor at termination is 24.

6.3 Spindle Mount Design

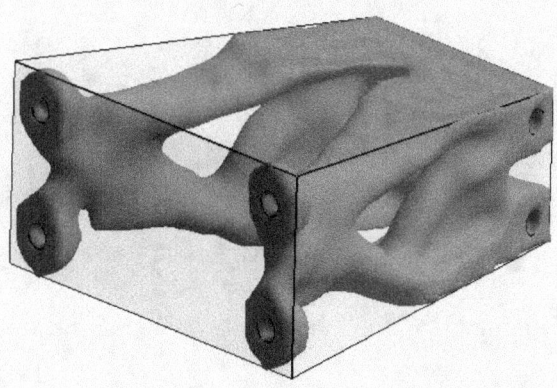

Figure 6.24: Optimized design for the spindle mount, with moment and torque neglected.

12. Save the project as "SpindleMountProjectSimplified.prj"

■ **Example 6.6 Spindle Mount Design Continued**
In the above example, we disregarded both the moment caused by the 500 lb force, and the 1300 lb-in torque. We will now include these, and compute the resulting forces on the four mounting holes. Towards this end, Figure 6.25 illustrates some of the critical dimensions (all dimensions are in inches) for the spindle clamp, and spindle mount.

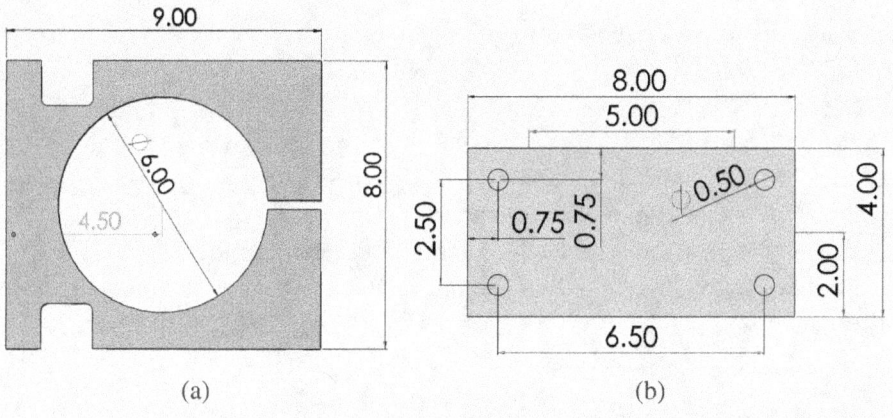

Figure 6.25: (a) Spindle clamp dimensions (inches), and (b) Spindle mount dimensions (inches).

Given these dimensions, using force and moment balance, one can compute the net force on the four mounting surfaces (see Figure 6.26).

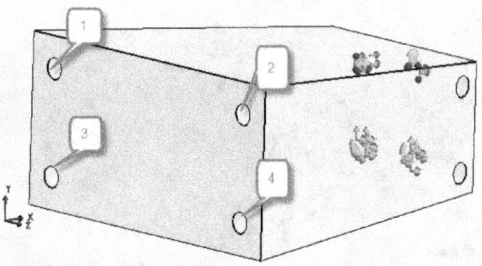

Figure 6.26: The four spindle mounting surfaces.

Specifically, observe that the moment caused by the 500 lb force is 500*4.5, i.e., 2250 lb-in. Since the vertical distance between the pair of holes is 2.5 in, this leads to the following forces on the four surfaces
1. 1: 450 lb along +x
2. 2: 450 lb along +x
3. 3: 450 lb along -x
4. 4: 450 lb along -x

Similarly, since the torque is 1300 lb-in, and the horizontal distance between the pair of holes is 6.5 in, this leads to the following forces:
1. 1: 100 lb along -x
2. 2: 100 lb along +x
3. 3: 100 lb along -x
4. 4: 100 lb along +x

In summary, the forces on the four mounting holes are:
1. 1: (350, 500, 0) lb
2. 2: (550, 500, 0) lb
3. 3: (-550, 500, 0) lb
4. 4: (-350, 500, 0) lb

We will now modify the spindle mount project.
1. Upload the previously saved SpindleMountProjectSimplified.prj.
2. Modify the forces on the four mounting holes to the ones computed above; see Figure 6.27.

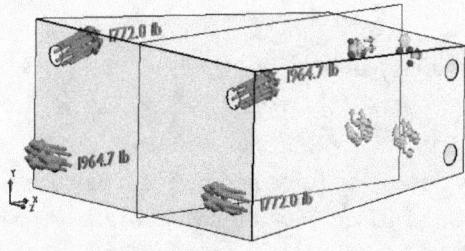

Figure 6.27: The forces on the four spindle mounting surfaces.

3. Save the project as "SpindleMountProject.prj"
4. Optimize the topology with the same set of constraints as before, for a desired volume fraction of 0.3. The resulting topology of 0.3 volume fraction is illustrated in Figure 6.28; the stress safety factor is around 9.

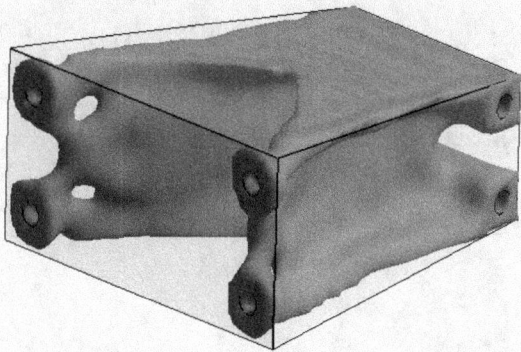

Figure 6.28: Optimized design for the spindle mount, with moment and torque included.

This example illustrates that incorrect set of forces can lead to a significantly different topology, and erroneous safety factors.

6.4 Exercises

Exercise 6.1 Set the units set to meters, and load CrossLink.stl from ParetoExamples folder; see Figure 6.29. Set the material to AlloySteel. Apply sliding boundary conditions on holes 1 and 2; apply a horizontal force (x-direction) of 2000 N on hole-3. Use a medium quality mesh, apply z-symmetry and carry out FEA. Under topopt constraints, select "Keep Fixed Faces", and apply. Now optimize for a desired volume fraction of 0.4. Note the topology and the final stress safety factor.

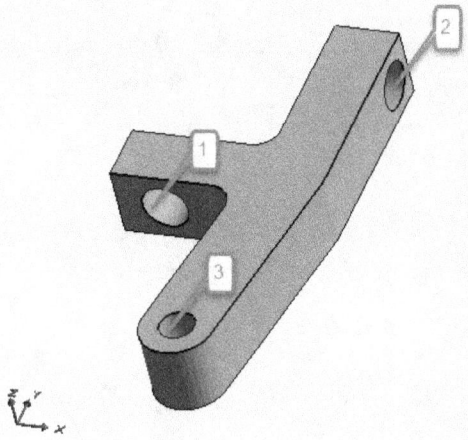

Figure 6.29: Cross link geometry.

Exercise 6.2 Repeat the CrossLink exercise above, with (a) hole-1 fixed, and hole-2 sliding, (b) hole-1 sliding, and hole-2 fixed, and finally (c) both holes fixed. Compare and contrast the topologies and safety factors.

Exercise 6.3 Consider the spindle mount problem; suppose there is an additional shear force of 300 lb along the z direction as illustrated in Figure 6.30. Compute the resulting force at the four mounting holes. Upload the previously saved SpindleMountProject.prj; modify the forces to include this additional force. Recompute the topology and safety factors. .

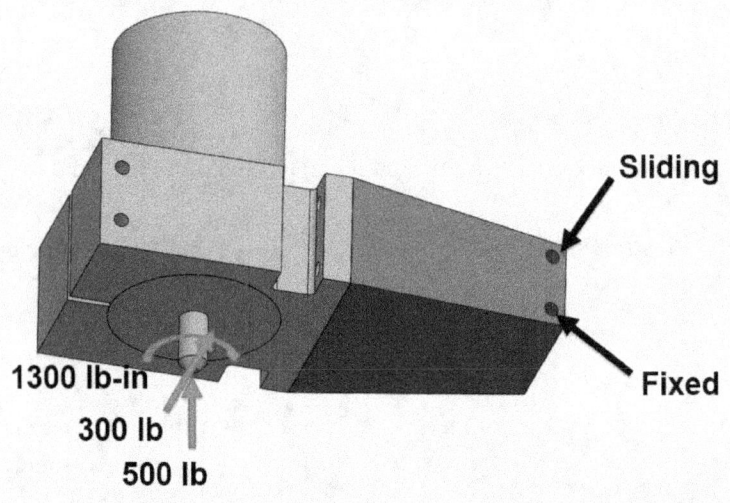

Figure 6.30: Spindle mount with an additional shear force.

Exercise 6.4 Upload the previously saved SpindleMountProject.prj. Switch the fixed and sliding boundary conditions; does the topology change?

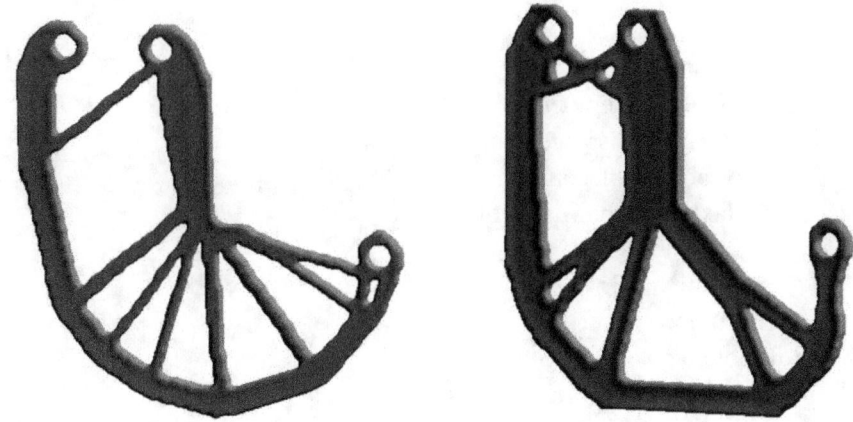

7. Performance Objectives & Constraints

In Chapter 4, we introduced the simplest and most common topology optimization problem, namely, compliance minimization subject to a volume constraint:

$$\begin{array}{l} \underset{\Omega \subset D}{minimize\, J} \\ \text{subject to} \\ V \leq V_0 \\ \mathbf{Ku} = \mathbf{f} \end{array} \qquad (7.1)$$

The solution to such problems leads to *stiff* designs of desired volume fraction. In this chapter, we will: (1) impose additional constraints such as displacement, stress and modal constraints, and (2) consider other objectives, besides compliance.

7.1 Performance Constraints

To begin with, let us consider the task of imposing additional constraints on the compliance minimization problem. Mathematically, one can impose constraints as follows:

$$\begin{array}{l} \underset{\Omega \subset D}{minimize\, J} \\ \text{subject to} \\ V \leq V_0 \\ \delta_{max} \leq \delta_0 \\ \sigma_{max} \leq \sigma_0 \\ \lambda_{min} \geq \lambda_0 \\ \mathbf{Ku} = \mathbf{f} \end{array} \qquad (7.2)$$

where:

J : Compliance
V_0 : Desired volume fraction
δ_{max} : Maximum displacement
δ_0 : Allowable displacement
σ_{max} : Maximum vonMises stress
σ_0 : Allowable stress
λ_{min} : Minimum (first) eigenvalue   (7.3)
λ_0 : Eigenmode lowerlimit
Ω : Topology to be computed
D : Domain within which the topology must lie
u : Finite element displacement field
K : Finite element stiffness matrix
f : External force vector

In the Pareto algorithm, the above problem is solved by tracing the compliance-volume fraction curve, and terminating the optimization algorithm when any of the constraints is violated. Alternately, one can consider the Lagrangian to incorporate the constraints, and minimize the total volume. However, pedagogical reasons, a simpler, but equally effective Pareto tracing algorithm was implemented. To illustrate, consider the following example.

■ Example 7.1 L-Bracket: Minimizing Compliance

1. Load the project *LBracketProject.prj*; the LBracket model is restrained at two locations, while a load of 5000N is applied at the tip; see Figure 7.1.

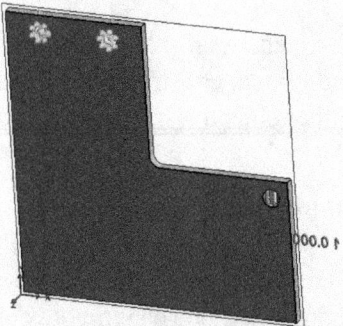

Figure 7.1: L-Bracket model subject to restraints and loads.

2. Carry out a static FEA with the default settings; the stress plot is illustrated in Figure 7.2a. Observe that the maximum displacement is around 0.00021 (m) and the maximum stress is around 176 MPa, as illustrated in Figure 7.2b.

| (a) | (b) |

Loadset: 0
#Elements: 49704
MeshSize: 0.00159
Deformation Scale: 49
Max Displacement: 0.00021 (meter)
Max Stress: 1.76e+08 (N/m^2)
Stress Safety Factor: 2.85
Time taken (sec): 3

Figure 7.2: (a) Initial stress plot, and (b) static FEA results for the initial design.

3. Next carry out a modal FEA with the default settings; the first eigenmode is illustrated in Figure 7.3a. Observe that the first mode is around 100 Hz as illustrated in Figure 7.3b.

EigenMode: 1, 100.29 (Hz)
#Elements: 49704
MeshSize: 0.00159
Deformation Scale: 49

| (a) | (b) |

Figure 7.3: (a) First mode, and (b) modal results for the initial design.

4. We will now impose the following optimization constraints (see Figure 7.4a):
   - A displacement constraint of 0.00406 (m).
   - A stress safety factor (constraint) of 1, i.e., the stress constraint is the yield strength of 500 MPa.
   - A lower limit of 90 Hz on the first mode.

   ⓡ A stress safety factor is imposed instead of a direct constraint on the stress for the following reason: in a multi-component assembly, different components may have different yield strengths, and imposing a direct stress constraint is not meaningful. Instead, a stress safety factor, with respect to the corresponding yield strength, is imposed across all components.

(R) Imposing a modal constraint entails carrying out a modal FEA at each step of the optimization process. This can be computationally expensive. Therefore, by default, the modal constraint is not imposed, i.e., the user must explicitly impose this constraint by providing a positive value for the lowerlimit of the 1st eigen-value. If a negative value is provided (default), then the modal constraint is not imposed.

(R) Further note that if a modal constraint is imposed, then the lower limit supplied must be less than the eigen-value of the initial design. Else, the optimization will terminate immediately since it will fail the constraint in the first step.

5. The objective is to minimize compliance, and the desired volume fraction is 0.5, as illustrated in Figure 7.4b.

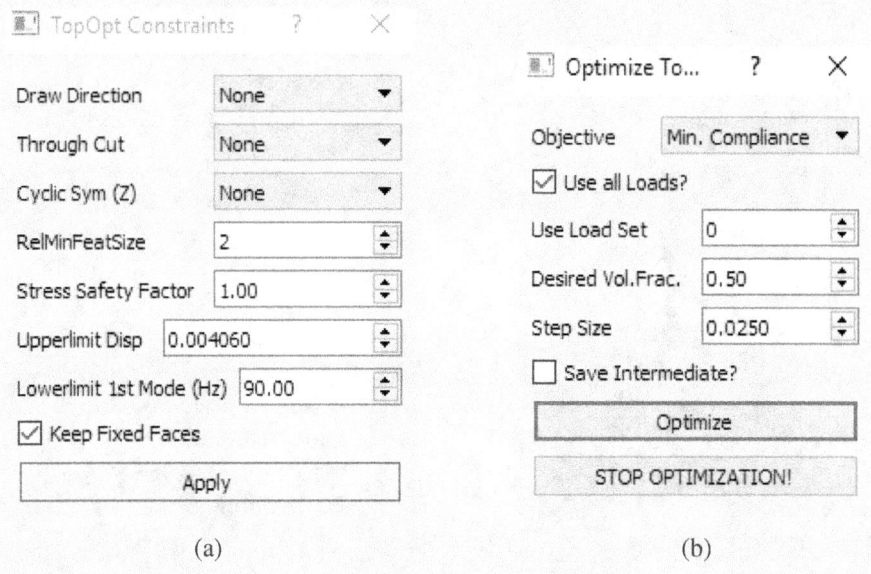

Figure 7.4: (a) Optimization constraints, and (b) objective and desired volume fraction.

6. Carry out topology optimization. The algorithm terminates at a volume fraction of 0.5 as illustrated in Figure 7.5a, while the safety factors for the optimized model are summarized in Figure 7.5b.

7.1 Performance Constraints

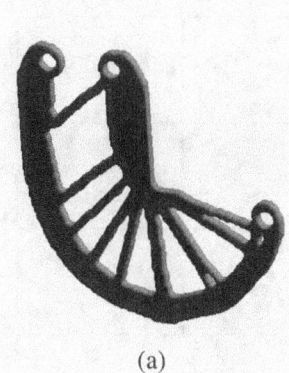

```
Target Vol. Frac: 0.500
Current Vol. Frac: 0.500
Time Taken: 200.3 (secs)
Estimated time: 121.9 (secs)
Deformation Scale: 1
Max Disp = 0.000266 (meter)
Disp. Safety Factor = 15.26
Max Stress = 2.2e+08 (N/m^2)
Stress Safety Factor = 2.27
First Mode = 114.42 (Hz)
Modal Safety Factor = 1.27
TopOpt complete!
```

(a)   (b)

Figure 7.5: (a) Volume constrained topology, and (b) safety factors for the optimized model.

> **R** Observe that all the safety factors are greater than 1, i.e., the displacement, the stress and modal constraints are not active. The algorithm terminated due to the desired volume fraction (constraint) of 0.5.

7. One can now plot the various quantities of interest as a function of volume fraction, using the TopOpt Results menu; see Figure 7.6.

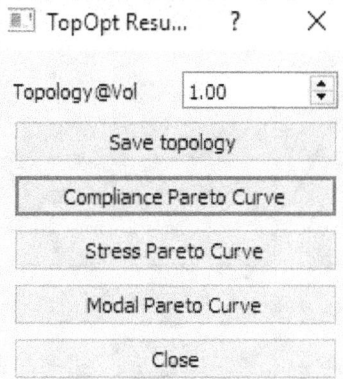

Figure 7.6: TopOpt Results menu.

8. The compliance curve is illustrated in Figure 7.7. It illustrates how the compliance changes as material is removed; observe that the compliance increases monotonically with decreasing volume fraction.

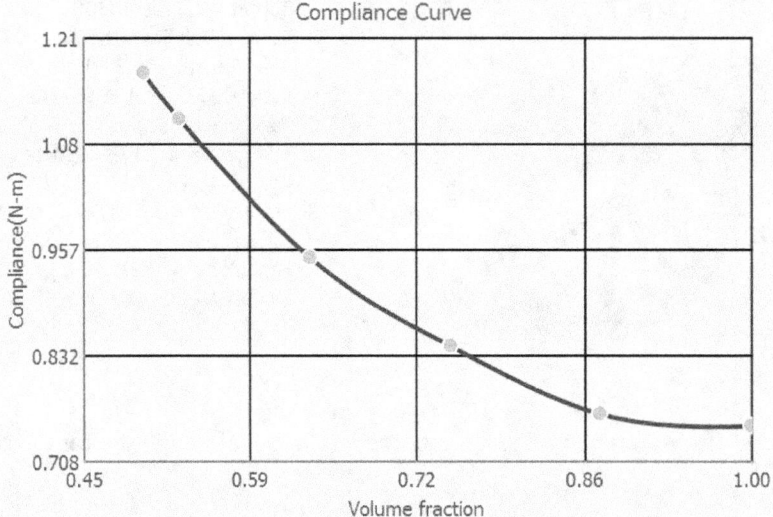

Figure 7.7: Compliance versus volume fraction curve.

9. Similarly, the maximum stress versus volume fraction is illustrated in Figure 7.8. Observe that the stress remains constant initially, and then increases with decreasing volume fraction.

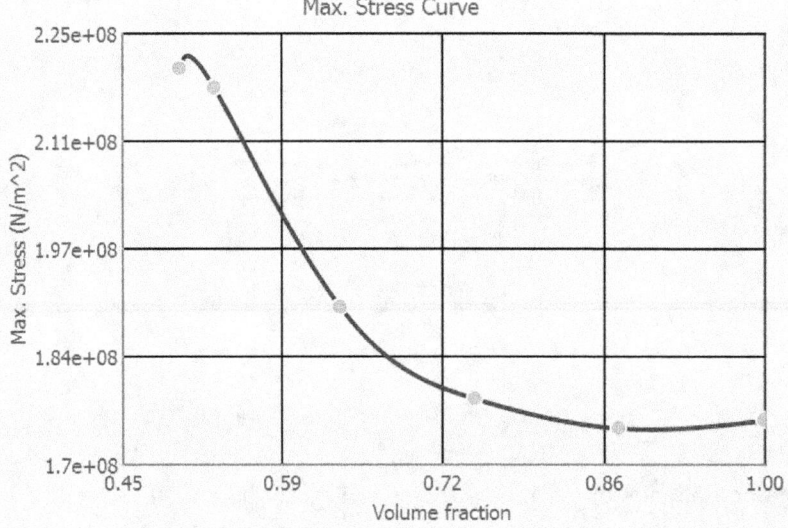

Figure 7.8: Max stress versus volume fraction curve.

10. Finally, the first eigenvalue curve is illustrated in Figure 7.9. Unlike the previous two curves, the eigenvalue does not exhibit a monotonic behavior, and this is very typical.

7.1 Performance Constraints

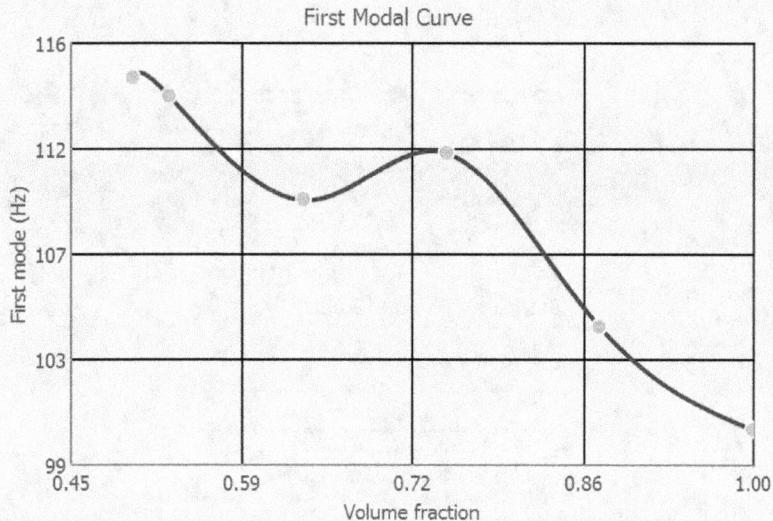

Figure 7.9: First eigenvalue versus volume fraction curve.

11. Next, we will lower the allowable displacement to 0.0004, as illustrated in Figure 7.10, and *Apply*.

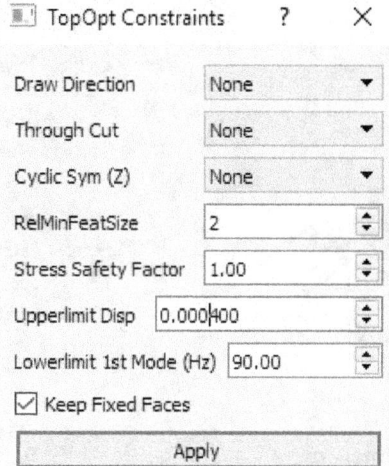

Figure 7.10: Changing the displacement limit to to 0.0004.

12. We will also lower the desired volume fraction to 0.01 as illustrated in Figure 7.11.

Chapter 7. Performance Objectives & Constraints

Figure 7.11: Changing the desired volume fraction to 0.01.

13. Repeat the optimization; the algorithm terminates at a volume fraction of 0.328 as illustrated in Figure 7.12a, while the safety factors are summarized in Figure 7.12b.

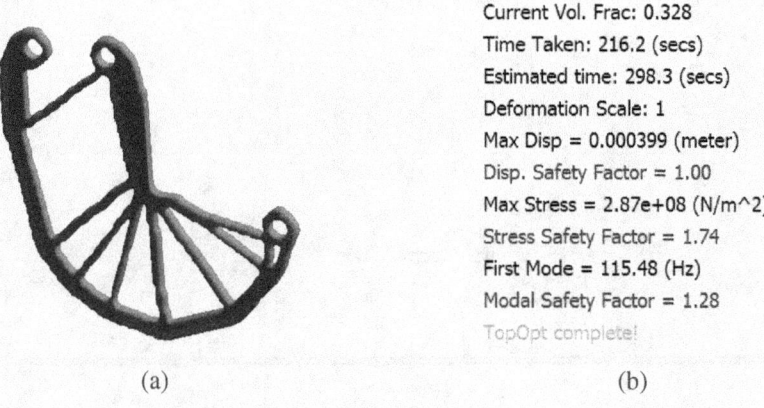

(a)   (b)

Figure 7.12: (a) Displacement constrained topology, and (b) safety factors.

R   Observe that the displacement safety factor is 1, i.e., the algorithm terminated due to the displacement constraint.

14. Now change the displacement constraint back to 0.00406, and increase the stress safety factor to 2 as in Figure 7.13.

7.1 Performance Constraints 87

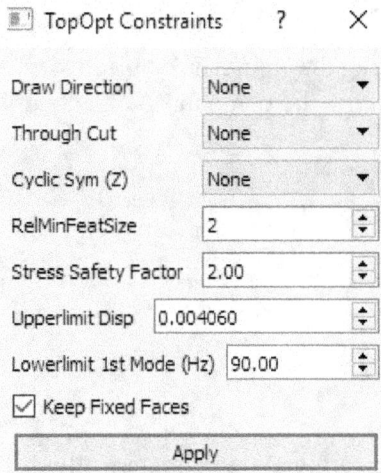

Figure 7.13: Changing the stress safety factor to 2.

15. Repeat the optimization; the algorithm terminates at a volume fraction of 0.377 as illustrated in Figure 7.14a, while the safety factors are summarized in Figure 7.14b. Observe that the algorithm terminated due to the stress constraint, i.e., the final stress safety factor is close to the desired value of 2.

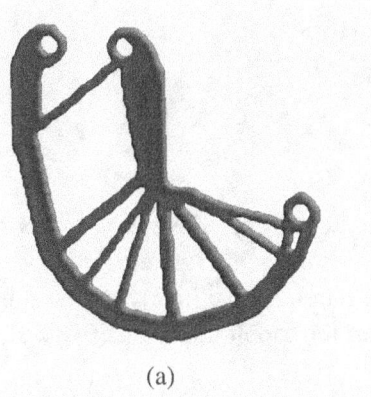

Target Vol. Frac: 0.010
Current Vol. Frac: 0.377
Time Taken: 232.4 (secs)
Estimated time: 254.9 (secs)
Deformation Scale: 1
Max Disp = 0.000347 (meter)
Disp. Safety Factor = 11.70
Max Stress = 2.49e+08 (N/m^2)
Stress Safety Factor = 2.01
First Mode = 115.41 (Hz)
Modal Safety Factor = 1.28
TopOpt complete!

(a)   (b)

Figure 7.14: (a) Displacement constrained topology, and (b) safety factors.

16. In compliance minimization, the first eigenvalue typically does not drop below the initial value. Therefore, the eigenvalue constraint will not play an important role.

The above example illustrates that, at termination, typically one of the constraints will be active. Further, since the objective was to minimize compliance, all the above designs are stiff designs.

7.2 Stress Minimization

Next we will consider replacing the compliance objective with a suitable measure of stress. For example, consider replacing the compliance objective with the maximum vonMises stresss:

$$\begin{aligned}
&\underset{\Omega \subset D}{\text{minimize}}\ \sigma_{max} \\
&\text{subject to} \\
&V \leq V_0 \\
&\delta_{max} \leq \delta_0 \\
&\sigma_{max} \leq \sigma_0 \\
&\lambda_{min} \geq \lambda_0 \\
&\mathbf{Ku} = \mathbf{f}
\end{aligned} \qquad (7.4)$$

Unfortunately, the above formulation is numerically difficult to solve due to the non-smooth nature of the maximum vonMises stress. We will therefore replace the maximum vonMises stress with a smooth measure of the stress:

$$\sigma^p = \left(\sum_e (\sigma_e)^p \right)^{1/p} \qquad (7.5)$$

where σ^p is referred to as the p-norm stress, while σ_e is the vonMises stress in element 'e'. The parameter p typically takes a value between 4 and 6; for large values of p, σ^p approaches σ_{max}, leading to the following problem:

$$\begin{aligned}
&\underset{\Omega \subset D}{\text{minimize}}\ \sigma^p \\
&\text{subject to} \\
&V \leq V_0 \\
&\delta_{max} \leq \delta_0 \\
&\sigma_{max} \leq \sigma_0 \\
&\lambda_{min} \geq \lambda_0 \\
&\mathbf{Ku} = \mathbf{f}
\end{aligned} \qquad (7.6)$$

The optimization algorithm will now generate 'strong' designs, as opposed to stiff designs, and will terminate when any of the constraints is violated. However, note that to solve the above problem, one must trace the Pareto curve involving the p-norm stress and volume fraction. This is done by computing the topological sensitivity of σ^p, and is discussed, for example, in:

- *Stress-constrained topology optimization: a topological level-set approach*, K. Suresh, M. Takalloozadeh, Struct Multidisc Optim 48, 295–309. doi:10.1007/s00158-013-0899-4, 2013.

To illustrate the impact of changing the objective, we will now revisit the L-bracket example.

■ **Example 7.2 L-Bracket: Minimizing Stress**

1. Once again load the LBracketProject.prj; see Figure 7.1 illustrated earlier.
2. The optimization constraints are as before (see Figure 7.4a):

7.2 Stress Minimization

- The displacement limit is set to 0.00406 (m).
- The stress limit is the yield strength of 500 MPa.
- The first eigen mode constraint is 90 Hz.

3. However, change the objective to Min. Stress, as illustrated in Figure 7.15.

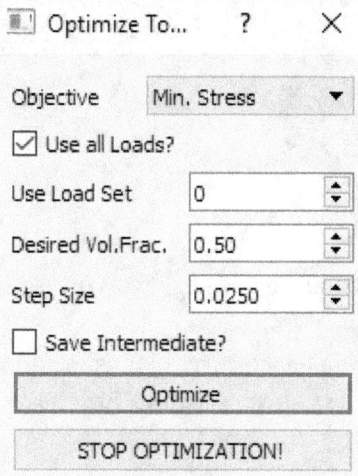

Figure 7.15: The objective is to minimize the p-norm stress.

4. Carry out the optimization; the algorithm terminates due to the imposed volume fraction constraint, and the optimized model of 0.5 volume fraction is illustrated in Figure 7.16a, while the associated safety factors are illustrated in Figure 7.16b.

> (R) Compare the topology in Figure 7.16a against the one in Figure 7.5a. Observe the significant difference in topology. Further, observe that the final displacement safety factor has decreased to 12.95, while the stress safety factor has increased to 2.75, but none of these constraints is active. The algorithm terminated when the desired volume fraction of 0.5 was reached.

Target Vol. Frac: 0.500
Current Vol. Frac: 0.500
Time Taken: 111.5 (secs)
Estimated time: 292.3 (secs)
Deformation Scale: 1
Max Disp = 0.000313 (meter)
Disp. Safety Factor = 12.95
Max Stress = 1.81e+08 (N/m^2)
Stress Safety Factor = 2.76
First Mode = 111.07 (Hz)
Modal Safety Factor = 1.23
TopOpt complete!

(a)   (b)

Figure 7.16: (a) Volume constrained topology with stress objective, and (b) safety factors.

5. The corresponding compliance curve is illustrated in Figure 7.17. Since the objective is to minimize stress, the compliance increases at a faster rate than in Figure 7.7.

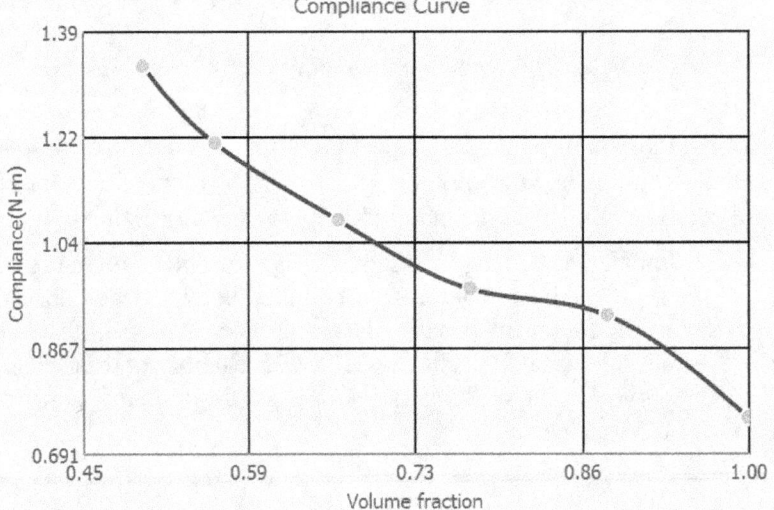

Figure 7.17: Compliance versus volume fraction.

6. The maximum stress curve is illustrated in Figure 7.18. Observe that the stress reduces initially, i.e., it is possible to remove material and reduce the maximum stress simulataneously! One can contrast this against Figure 7.8.

7.2 Stress Minimization

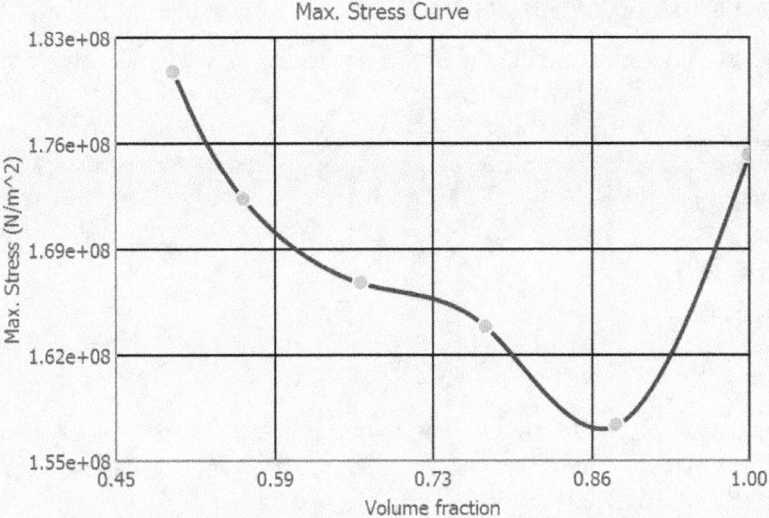

Figure 7.18: Maximum stress versus volume fraction.

7. The first eigenvalue curve is illustrated in Figure 7.19. Observe that the first eigenvalue also reduces initially (not desirable).

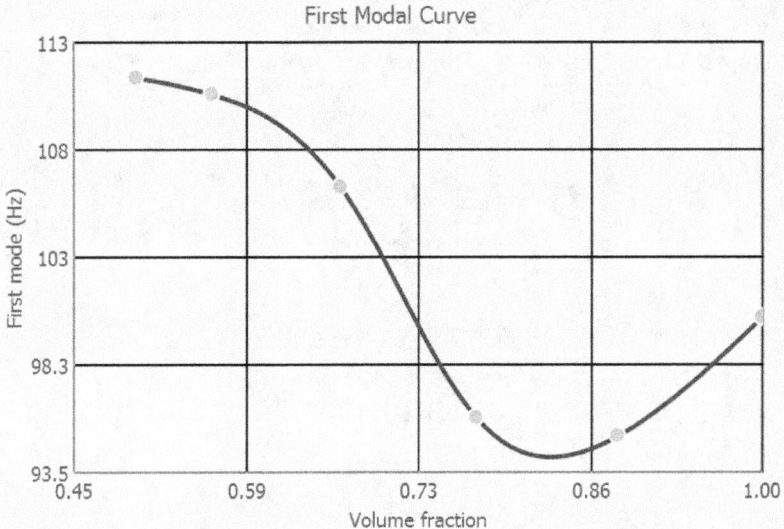

Figure 7.19: First eigenvalue versus volume fraction.

> The reader is encouraged to change the desired volume fraction and the constraints to better understand the interplay between these parameters.

7.3 Eigenvalue Maximization

Next we will consider maximizing the first eigenvalue, i.e., we will consider the problem:

$$\begin{aligned}
&\underset{\Omega \subset D}{\text{Maximize}} \; \lambda_{min} \\
&\text{subject to} \\
&V \leq V_0 \\
&\delta_{max} \leq \delta_0 \\
&\sigma_{max} \leq \sigma_0 \\
&\lambda_{min} \geq \lambda_0 \\
&\mathbf{Ku} = \mathbf{f}
\end{aligned} \tag{7.7}$$

In other words, we will trace the Pareto curve involving the first eigenvalue and the volume fraction. The topological sensitivity of the first eigenvalue can be determined using the methods discussed in references mentioned earlier.

■ **Example 7.3 L-Bracket: Maximizing First Eigenvalue**

1. Once again load the LBracketProject.prj; see Figure 7.1 illustrated earlier.
2. The optimization constraints are as before (see Figure 7.4a):
   - The displacement limit is set to 0.00406 (m).
   - The stress limit is the yield strength of 500 MPa.
   - The first eigenvalue constraint is 90 Hz.
3. Change the objective to Max Eigenvalue, as illustrated in Figure 7.20.

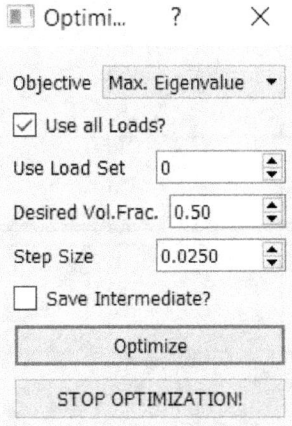

Figure 7.20: The objective is to maximize the first eigenvalue.

4. Carry out the optimization; once again, the algorithm terminates due to the imposed volume fraction constraint. The optimized model of 0.5 volume fraction is illustrated in Figure 7.21a, while the associated safety factors are illustrated in Figure 7.21b. Observe the significant difference in topology compared to the earlier compliance and stress based topologies.

7.3 Eigenvalue Maximization

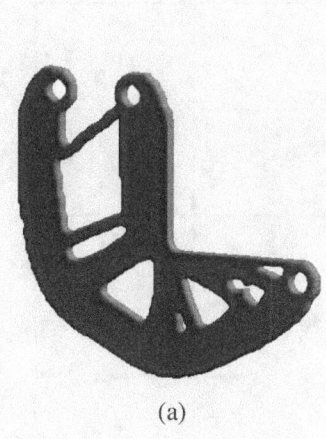

Target Vol. Frac: 0.500
Current Vol. Frac: 0.500
Time Taken: 164.8 (secs)
Estimated time: 290.6 (secs)
Deformation Scale: 1
Max Disp = 0.000281 (meter)
Disp. Safety Factor = 14.44
Max Stress = 2.46e+08 (N/m^2)
Stress Safety Factor = 2.03
First Mode = 142.67 (Hz)
Modal Safety Factor = 1.59
TopOpt complete!

(a)   (b)

Figure 7.21: (a) Volume constrained topology with eigenvalue objective, and (b) safety factors.

5. The corresponding compliance curve is illustrated in Figure 7.22. Unlike the previous examples, the compliance curve displays a non-monotonic behavior.

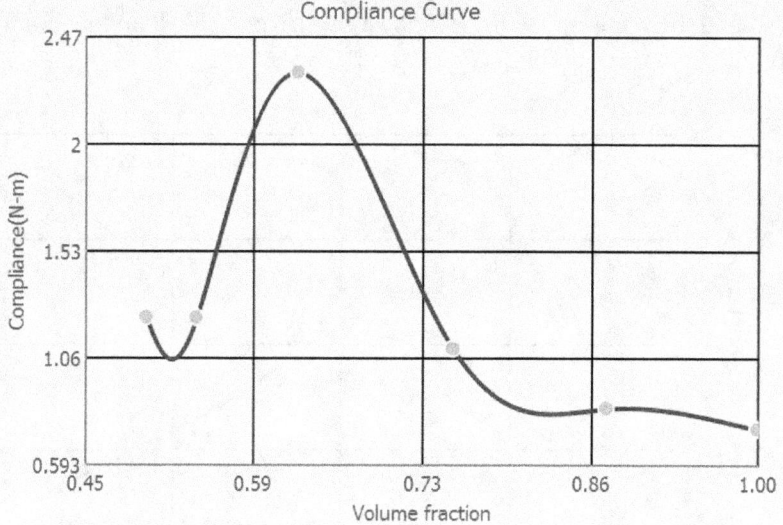

Figure 7.22: Compliance versus volume fraction.

6. The maximum stress curve is illustrated in Figure 7.23.

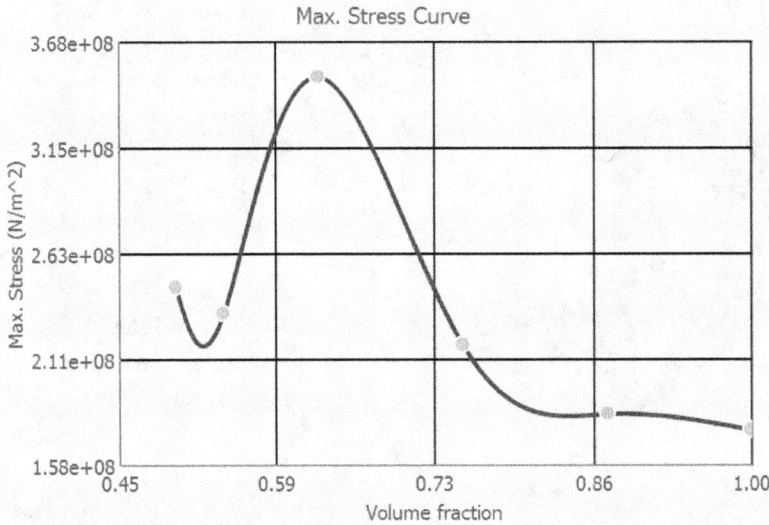

Figure 7.23: Maximum stress versus volume fraction.

7. The first eigenvalue curve is illustrated in Figure 7.24. Observe that the first eigenvalue increases significantly (desirable), and then decreases. If the objective is to find the topology with the highest first eigenvalue, then the graph suggests a volume fraction of around 0.6.

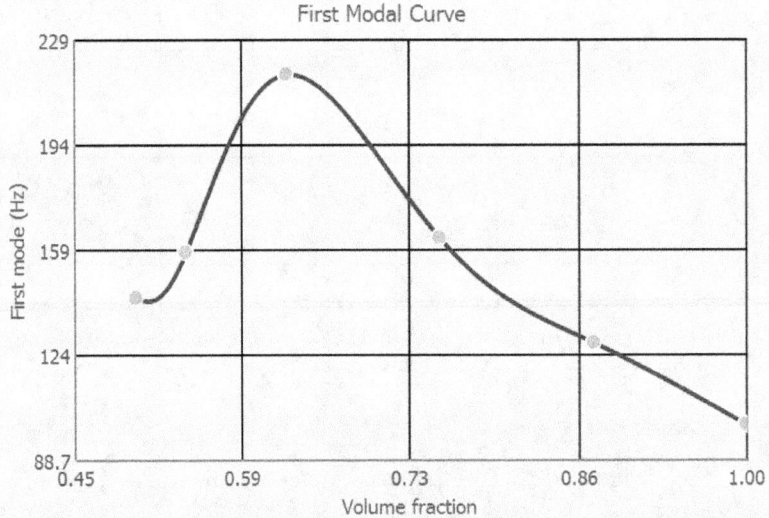

Figure 7.24: First eigenvalue versus volume fraction.

(R) The reader is encouraged to change the desired volume fraction and the constraints to better understand the interplay between these parameters.

7.4 Additional Examples

We will now illustrate the interplay between objectives and constraints through additional examples.

■ **Example 7.4 Split Bar**

1. Load the SplitBarProject.prj; see Figure 7.25.

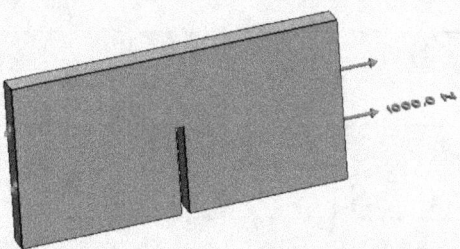

Figure 7.25: Split bar model subject to restraints and loads.

2. The optimization constraints are:
   - The displacement limit is set to 0.0008 (m).
   - The stress limit is the yield strength of 500 MPa.
   - No constraint is imposed on the eigenvalue.
3. With the objective as compliance, and desired volume fraction of 0.01, carry out the optimization.

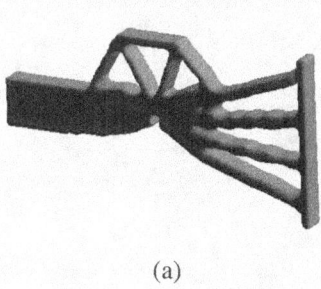

Target Vol. Frac: 0.010
Current Vol. Frac: 0.355
Time Taken: 32.5 (secs)
Estimated time: 84.0 (secs)
Deformation Scale: 1
Max Disp = 7.54e-05 (meter)
Disp. Safety Factor = 10.61
Max Stress = 4.99e+08 (N/m^2)
Stress Safety Factor = 1.00
TopOpt complete!

(a)   (b)

Figure 7.26: (a) Final topology with compliance objective, and (b) safety factors.

The optimized model of 0.348 volume fraction is illustrated in Figure 7.26a, while the associated safety factors are illustrated in Figure 7.26b.

® Observe that, at termination, the stress safety factor is 1, i.e., the optimization is stress constrained.

4. The compliance curve is illustrated in Figure 7.27. Observe that the compliance monotonically increases with decreasing volume fraction.

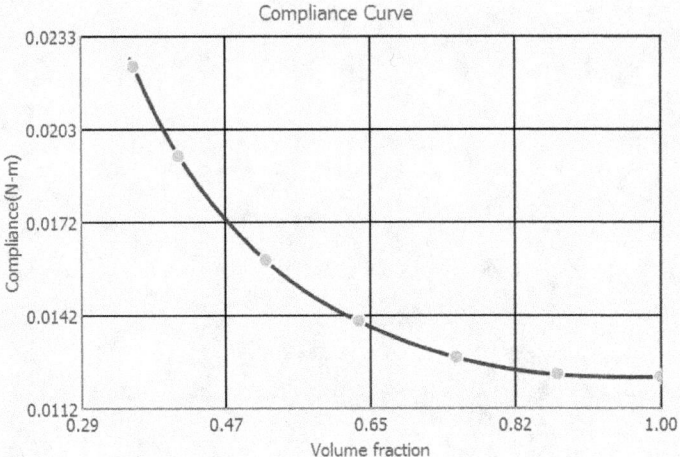

Figure 7.27: Compliance curve.

5. The maximum stress curve is illustrated in Figure 7.28. Observe that the stress remains constant initially, and then increases with decreasing volume fraction.

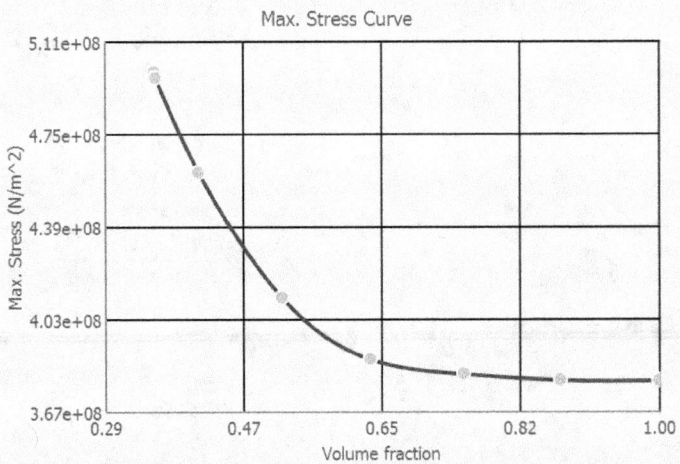

Figure 7.28: Maximum stress curve.

6. Next, change the optimization objective to Min Stress. Carry out the optimization; the final topology of 0.259 volume fraction is illustrated in Figure 7.29a, while the safety factors are summarized in Figure 7.29b.

7.4 Additional Examples

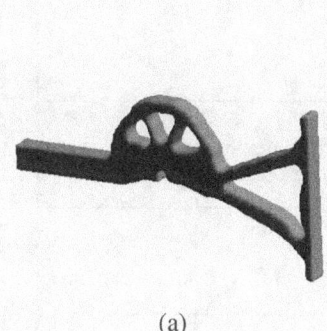

Target Vol. Frac: 0.010
Current Vol. Frac: 0.287
Time Taken: 40.9 (secs)
Estimated time: 94.6 (secs)
Deformation Scale: 1
Max Disp = 9.74e-05 (meter)
Disp. Safety Factor = 8.22
Max Stress = 4.55e+08 (N/m^2)
Stress Safety Factor = 1.10
TopOpt complete!

(a) (b)

Figure 7.29: (a) Final topology with stress objective, and (b) safety factors.

(R) Observe that the optimization is once again constrained by stress. However, we were able to achieve a lower volume fraction.

7. The compliance curve is illustrated in Figure 7.30. Observe that the compliance monotonically increases with decreasing volume fraction.

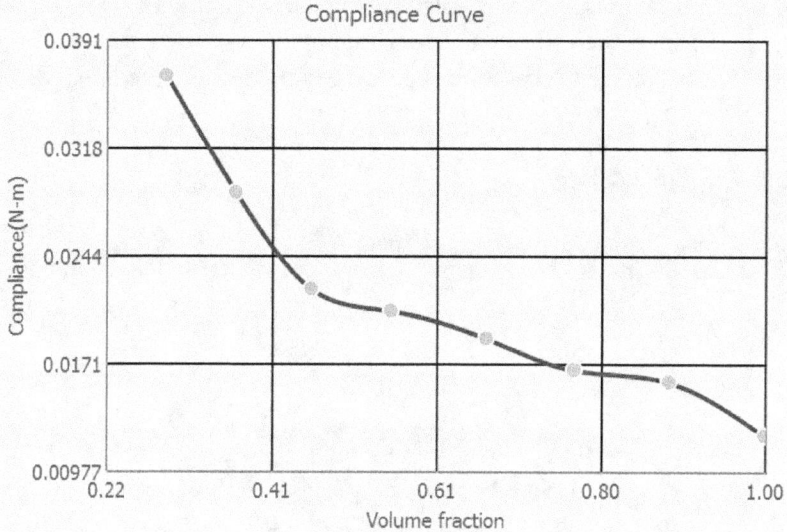

Figure 7.30: Compliance curve.

8. The maximum stress curve is illustrated in Figure 7.31. Observe that, the stress reduces constant initially, and then increases with decreasing volume fraction.

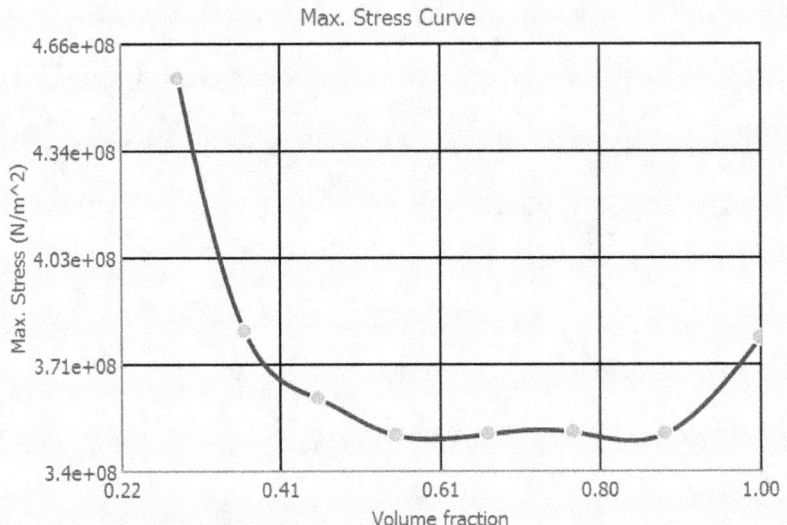

Figure 7.31: Maximum stress curve.

In the previous examples, the initial design exhibited a reentrant corner, i.e., a local stress raiser. In the presence of such stress raisers, changing the objective to Min Stress, will often lower the stress initially. We will now consider an example where the initial design does not exhibit a entrant corner.

■ **Example 7.5 Thick Plate**

1. Load the ThickPlate.prj; see Figure 8.14.

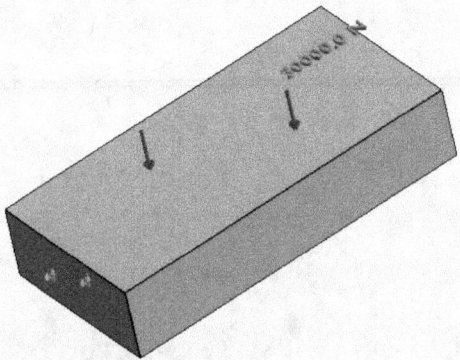

Figure 7.32: Thick plate model subject to restraints and loads.

2. The optimization constraints are:
    - The displacement limit is set to 2 mm.

7.4 Additional Examples

- The stress limit is the yield strength of 110 MPa.
- No constraint is imposed on the eigenvalue.

3. With the objective as compliance, and desired volume fraction of 0.01, carry out the optimization. The optimized model of 0.498 volume fraction is illustrated in Figure 7.33a, while the associated safety factors are illustrated in Figure 7.33b.

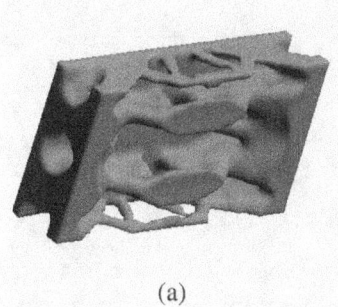

Target Vol. Frac: 0.010
Current Vol. Frac: 0.537
Time Taken: 22.1 (secs)
Estimated time: 51.1 (secs)
Deformation Scale: 1
Max Disp = 0.0753 (mm)
Disp. Safety Factor = 26.56
Max Stress = 102 (N/mm^2)
Stress Safety Factor = 1.08
TopOpt complete!

(a)   (b)

Figure 7.33: (a) Final topology with compliance objective, and (b) safety factors.

> (R) Observe that, at termination, the stress safety factor is close to 1, i.e., the optimization is stress constrained.

4. The compliance curve is illustrated in Figure 7.34.

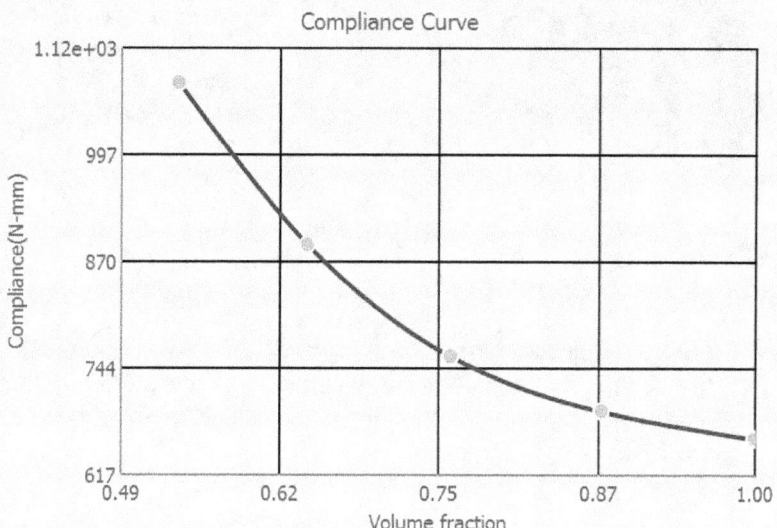

Figure 7.34: Compliance curve.

5. The maximum stress curve is illustrated in Figure 7.35.

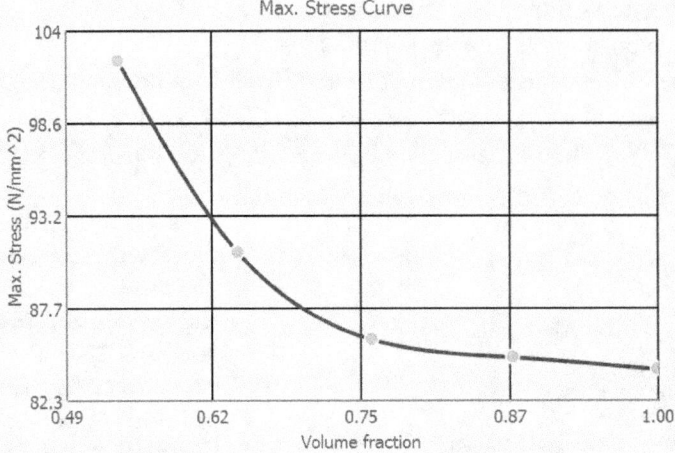

Figure 7.35: Maximum stress curve.

6. Next, change the optimization objective to Min Stress. Carry out the optimization; the final topology of 0.512 volume fraction is illustrated in Figure 8.16a, while the safety factors are summarized in Figure 8.16b.

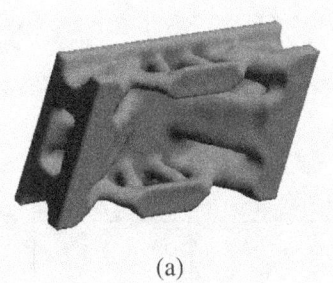

(a)

Target Vol. Frac: 0.010
Current Vol. Frac: 0.516
Time Taken: 42.8 (secs)
Estimated time: 103.7 (secs)
Deformation Scale: 1
Max Disp = 0.0839 (mm)
Disp. Safety Factor = 23.85
Max Stress = 109 (N/mm^2)
Stress Safety Factor = 1.00
TopOpt complete!

(b)

Figure 7.36: (a) Final topology with stress objective, and (b) safety factors.

(R) Observe that the optimization is once again constrained by stress. Further, the final topology and volume fraction is almost identical to the one associated with compliance objective. In other words,

7. The compliance curve is illustrated in Figure 7.37.

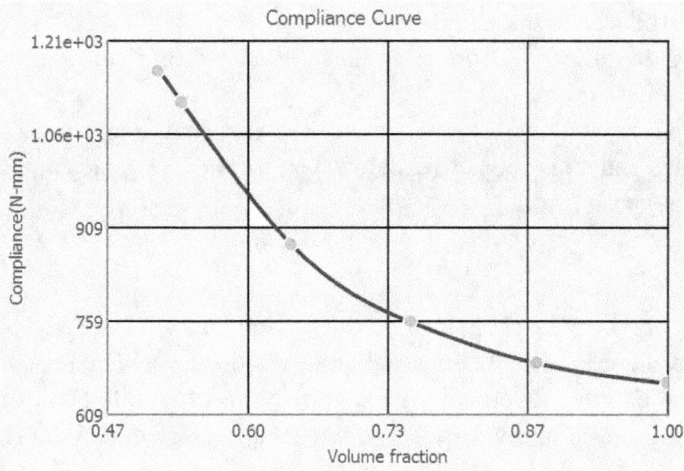

Figure 7.37: Compliance curve.

8. The maximum stress curve is illustrated in Figure 7.38. Observe that, the stress is more or less a constant initially, and then increases with decreasing volume fraction.

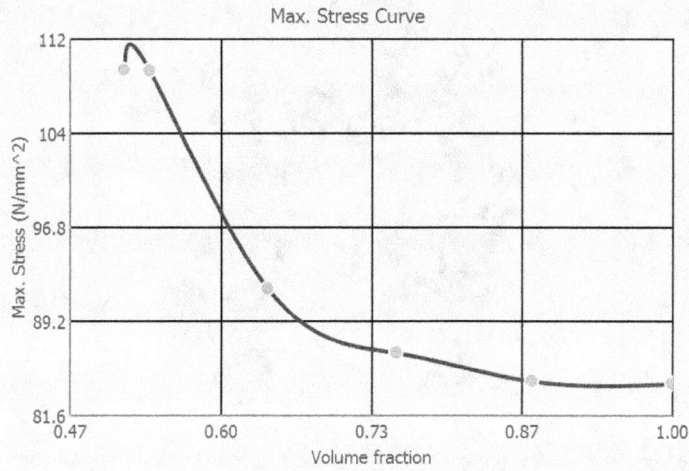

Figure 7.38: Maximum stress curve.

7.5 Conclusions

The main conclusions to draw from this chapter are:
1. Depending on the objective, the algorithm traces the Pareto curve involving the corresponding objective and volume fraction.
2. The algorithm will terminate when any of the imposed constraints is violated.
3. Changing the objective can significantly change the topology.
4. If the initial design has reentrant corners, with the stress objective, a large fillet radius

will be introduced at these corners.
5. Changing the objective from compliance to stress or eigenvalue will increase the computational time.
6. The compliance will increase monotonically with decreasing volume fraction, for both compliance and stress objectives. However, for the latter, the compliance will often increase at a faster rate.

7.6 Exercises

Exercise 7.1 Set the units set to meters (MKS), and load Knuckle.stl from ParetoExamples folder; see Figure 7.39. Set the material to Al1060. Apply fixed boundary conditions on holes 1 and 2; apply a vertical force (y-direction) of 10,000 N on surface-3. Use a medium quality mesh, apply x and z symmetry and carry out FEA. The stress safety factor should be around 2. Next, carry out modal analysis; the first eigenvalue should be around 2300 Hz. Under topopt constraints, select "Lowerlimit 1st Mode" and set the value to 2000 Hz, select "Keep Fixed Faces", and apply. Now optimize with compliance as objective, for a desired volume fraction of 0.01. Note the final volume fraction, topology and the safety factors. Plot the compliance, stress and eigenvalue curves.

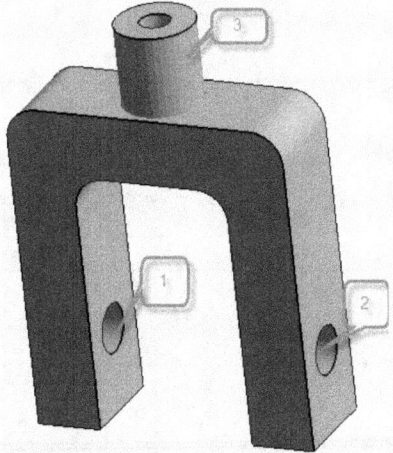

Figure 7.39: Knuckle geometry.

Exercise 7.2 Repeat the above optimization problem, but with stress as objective. Compare and comment on the results; plot the three curves.

Exercise 7.3 Repeat the above optimization problem, but with eigenvalue as objective. Compare and comment on the results; plot the three curves.

7.6 Exercises

Exercise 7.4 Consider the spindle mount example discussed in the previous chapter. Compute the first eigenvalue. Carry out the optimization as before, but with an additional constraint that the eigenvalue should not drop below 90 percent of the initial value. Carry out the optimization for all three objectives. Include the moment and torque in your optimization.

8. Design & Manufacturing Constraints

Designs generated through topology optimization are free-form and are often geometrically complex. This can pose manufacturing challenges. Further, engineers may be interested in imposing additional *design* constraints for aesthetics. In this chapter, we will discuss how these manufacturing and design constraints may be imposed; such constraints include

1. **Minimum feature size**: to ensure that the thickness of each geometric feature is larger than a prescribed value.
2. **Retained surfaces**: to prevent material from being removed from specified surfaces.
3. **Draw direction constraint**: to ensure that there no cavities along a given direction.
4. **Through cut**: to ensure that the cross-section remains constant along a given direction, i.e., to generate a 2.5D design.
5. **Cyclic symmetry**: to impose cyclic symmetry for circular geometries.

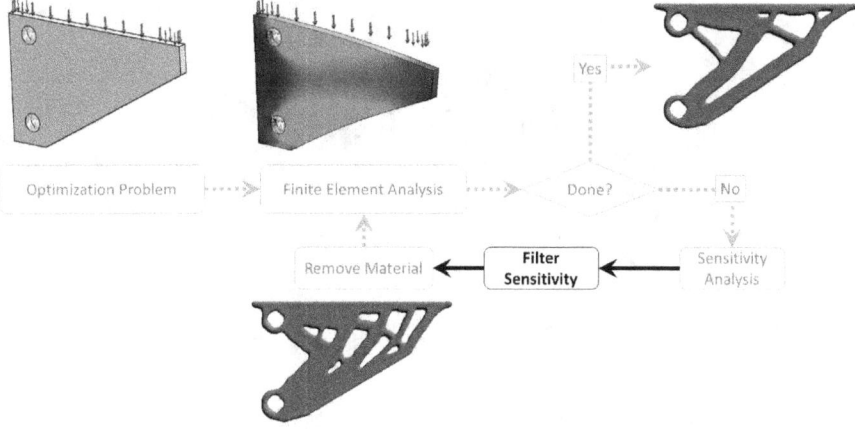

Figure 8.1: Design and manufacturing constraints are imposed through different filtering schemes.

To understand how these constraints may be imposed, recall the topology optimization workflow illustrated in Figure 8.1. As mentioned earlier, the sensitivity field dictates where material must be removed. Therefore, to modify the material removal process, one must *modify the sensitivity field*. These modifications are achieved by designing suitable *filters*. Thus, for each constraint, we must modify the sensitivity field; this is discussed and demonstrated below through examples.

8.1 Draw-Direction Constraint

The first manufacturing constraint we will consider is the *draw-direction constraint*. Imposing this constraint will ensure that there would be no cavities along the prescribed direction. This can be particularly useful for casting.

To illustrate the underlying concept, let us assume that we have arrived at a topology in Figure 8.2 (this example will discussed later in this section). Observe that this design is not well suited for casting since there are voids along the casting direction.

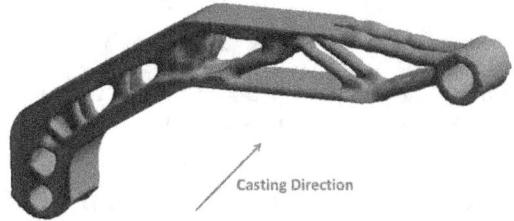

Figure 8.2: Without considering draw-direction constraint, the optimized design is not well suited for casting.

Before one can impose the draw-direction (casting) constraint, it is important to understand why the topology exhibits voids in the draw-direction. Towards this end, consider a single ray illustrated in Figure 8.3.

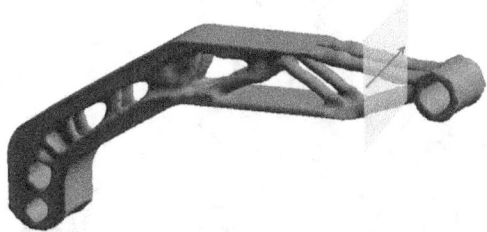

Figure 8.3: A cross-section that needs to be modified for casting.

A plausible topological sensitivity field along this ray-direction is illustrated in Figure 8.4. Observe that, given the sensitivity field, and a hypothetical cut-off value of 0.5, only the material with sensitivity field above 0.5 is retained, resulting in the topology of Figure 8.3. Thus, since the sensitivity field exhibits a valley, it leads to a void in the optimized design.

8.1 Draw-Direction Constraint

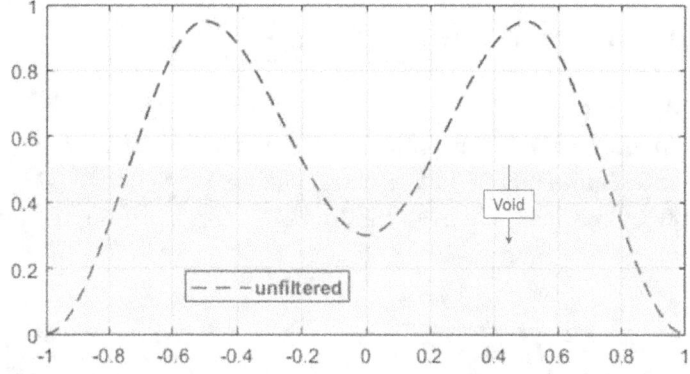

Figure 8.4: Unfiltered sensitivity field at the cross-section along casting direction.

To prevent voids, one must remove valleys in the sensitivity field as illustrated in Figure 8.5. The cutoff value for the sensitivity field must be appropriately modified to ensure the correct volume fraction is recovered.

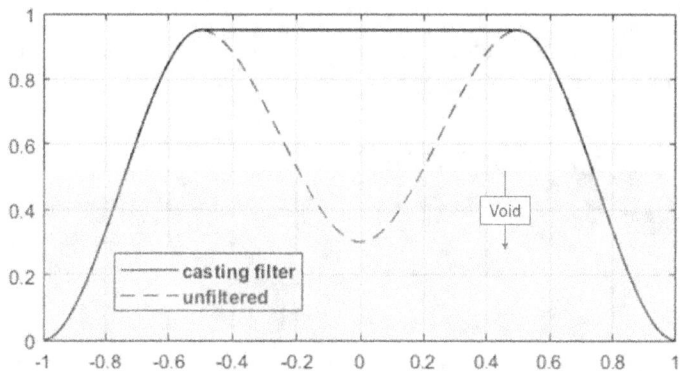

Figure 8.5: Effect of casting constraint on sensitivity field.

With the filtering implemented, the resulting optimized topology is shown in Figure 8.6. This illustrates the basic principle behind imposing the draw-direction constraint. A few examples are provided below to demonstrate this principle.

Figure 8.6: Filtered optimized design that can be cast.

▪ Example 8.1 Swing Arm Project

1. Apply MKS units.
2. Load *SwingArm.stl* geometry from the *ParetoExamples* folder
3. Select the default material of *Alloy Steel* and apply.
4. Select the cylindrical hole as in Figure 8.7a. Then, restrain the face in all three directions; see in Figure 8.7b.

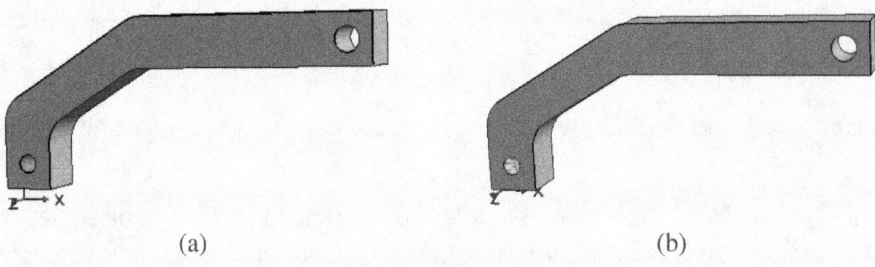

Figure 8.7: (a) Face selection, (b) restraint condition.

5. Select the face as illustrated in Figure 8.8a, and apply a vertical force of 1,500 N, as illustrated in Figure 8.8(b).

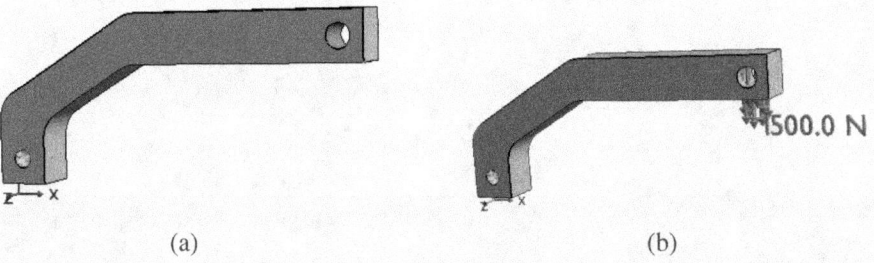

Figure 8.8: (a) Face selection and (b) applying load of 1,500 (N).

6. Set mesh quality to *Fine*, this should change number of elements to 100,000.
7. The *Z Symmetry* option is selected; this implies that the algorithm will attempt to maintain symmetry along z-direction during meshing (and later during optimization).
8. Carry out an FEA (Figure 8.9a). The maximum stress is 307 MPa.

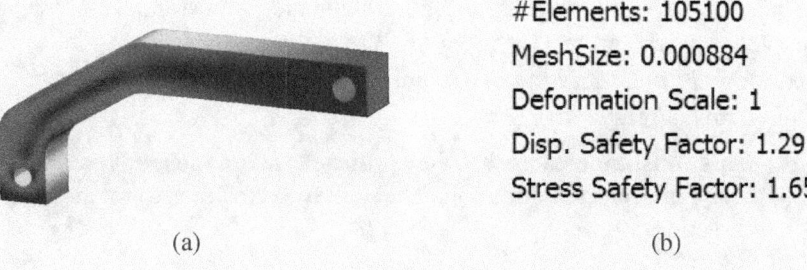

```
#Elements: 105100
MeshSize: 0.000884
Deformation Scale: 1
Disp. Safety Factor: 1.29
Stress Safety Factor: 1.65
```

(a)   (b)

Figure 8.9: (a) Stress results, and (b) safety factors.

(R) For long slender bodies, a high-quality mesh is usually recommended.

(R) Based on the yield strength for alloy steel, we observe that the stress safety factor is 1.63; see Figure 8.9b (we will disregard the displacement safety factor).

9. We will now carry out topology optimization using the following parameters:
   - The desired volume fraction is set to 0.1.
   - Displacement limit is set to 0.0016 (m).
   - The *Keep Fixed Faces* is selected; this means that material surrounding the two restrained holes will be retained during optimization
10. The optimization terminates at a volume fraction of 0.5, and the optimized model is illustrated in Figure 8.10b.

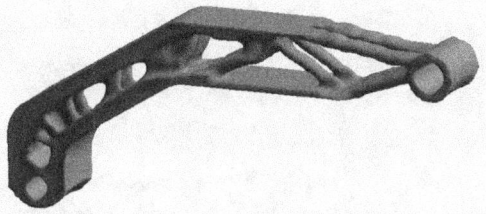

Figure 8.10: Optimized model at 0.5 volume fraction.

11. The maximum stress is 308 MPa, and stress safety factor at termination is 1.63. Due to the presence of thin features, the optimization usually terminates early.

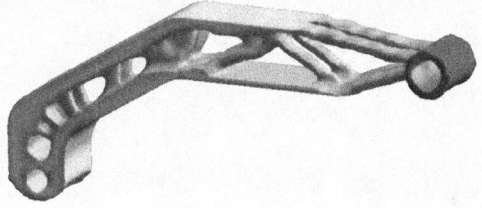

Figure 8.11: Final stress distribution of optimized swing arm.

12. You can now save the current project, i.e. the material, restraints and loads. When prompted, save the project as *SwingArmProject.prj*.
13. Next, we will impose casting constraint.
14. Set the *Draw Direction* to *ZDir*.

    ⓡ This constraint states that the optimized model must not have voids along the selected direction. This will force features to merge in that direction.

15. The optimization terminates at a volume fraction of 0.36, and the optimized model is illustrated in Figure 8.12.

    ⓡ Observe that the model does not have voids along the z direction. Furthermore, we could achieve a lower volume fraction of 0.37, with a stress safety factor of 1.2.

Figure 8.12: Optimized swing arm with draw direction constraint along Z at volume fraction of 0.36.

16. One can view the stresses on the optimized topology by changing the display mode (see Figure 8.13).

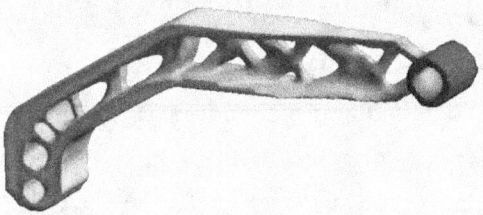

Figure 8.13: Final stress distribution of optimized swing arm with draw direction along Z.

■ **Example 8.2 Thick plate project**

1. Change the units to mmKS.
2. Load *ThickPlate.stl* (since the stl file was saved in mm), as in Figure 8.14.

8.1 Draw-Direction Constraint

Figure 8.14: Thick plate model (mm units).

3. Apply *CastIronGr25* material.
4. Change display to transparent.
5. Restrain the two side surfaces in all directions as in Figure 8.15a.
6. Next, apply a load of 30,000 N as in Figure 8.15b.

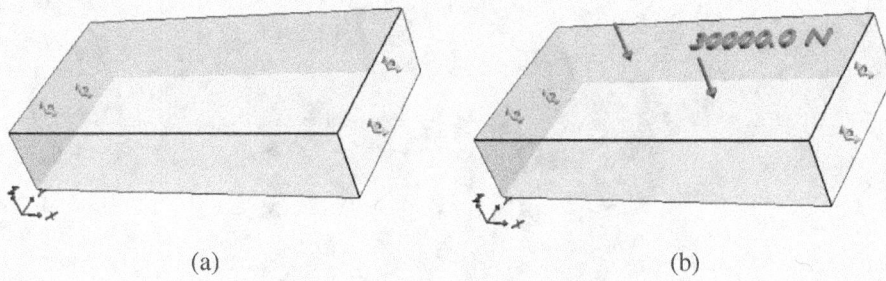

(a) (b)

Figure 8.15: (a)Restraining the two side faces, and (b) applying a normal force of 30,000 N.

7. Impose symmetry along X and Y directions.
8. The results of static analysis with 50,000 elements is illustrated in Figure 8.16. The maximum stress is 84 N/mm^2, i.e., 84 MPa.

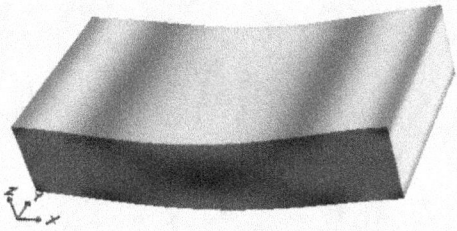

Figure 8.16: Finite element stress result.

9. Now optimize using the following parameters:
    - Desired volume fraction at 0.01.
    - Set *RelMinFeatSize* to 2.

- Set *Stress Safety Factor* to 1.00.
- Set *Displacement Limit* to 2.

10. The final optimized design (0.45 volume fraction) is illustrated in Figure 8.17b.

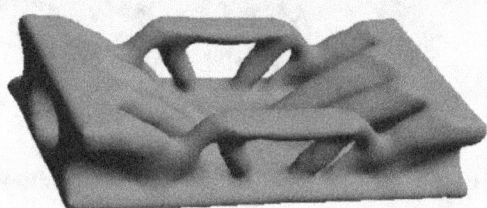

Figure 8.17: Optimized ThickPlate at 0.45 volume fraction.

11. We will now impose a casting constraint. Under *TopOpt Constraints*, set *DrawDirection* option to z-direction.
12. The optimized model (at a volume fraction of 0.45) is illustrated in Figure 8.18.

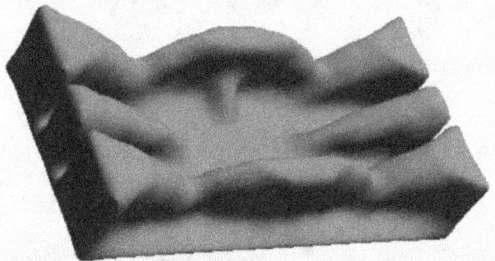

Figure 8.18: Optimized ThickPlate with draw direction constraint along Z at volume fraction of 0.45.

> (R) Observe the simplicity of the design, that additional *cores* may not be needed for casting, and the surfaces are accessible for machining.

8.2 Through-Cut Constraint

The through-cut constraint is similar to the draw-direction constraint in that a direction must be chosen, but the cross section will remain constant along a given direction, to generate 2.5D designs. Such designs are amenable to various manufacturing processes, such as laser cutting, or LENS 3D metal printing. In this section, we will introduce the basic concept behind the through-cut constraint to achieve this goal.

Consider again the optimized swing arm in Figure 8.19. Observe that the cross-section is not a constant along the through-cut direction.

8.2 Through-Cut Constraint

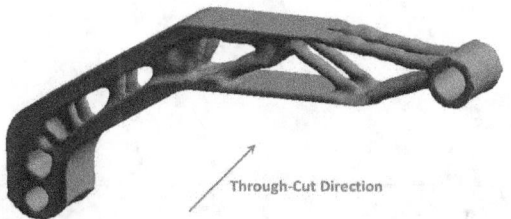

Figure 8.19: Without considering through-cut constraint, the optimized design cannot be manufactured by laser cutting.

To impose the through-cut constraint, consider a single ray indicated in Figure 8.20.

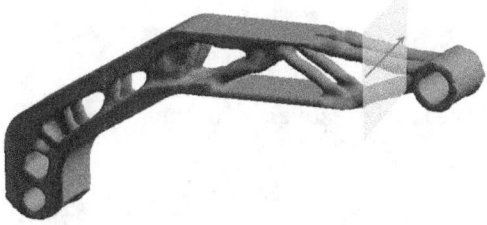

Figure 8.20: A cross-section that needs to be modified for laser cutting.

A plausible sensitivity field along this ray direction is illustrated in Figure 8.21.

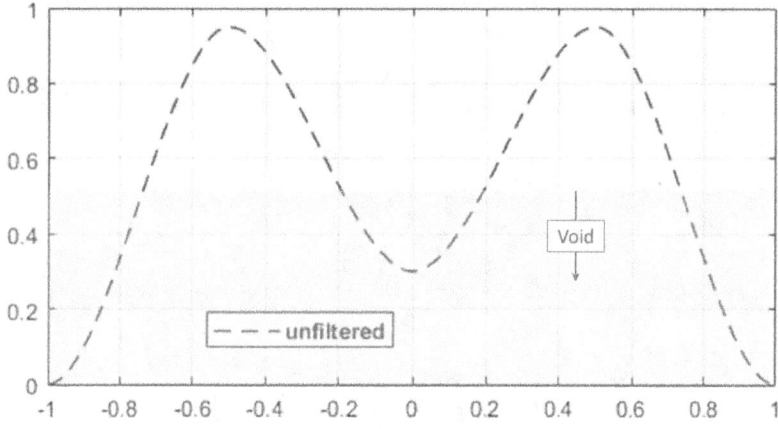

Figure 8.21: Unfiltered sensitivity field at the cross-section along through-cut direction.

To impose the through-cut constraint, we must ensure that the sensitivity values along through-cut direction are exactly the same (so that material along that ray would either be retained enitrely, or removed entirely). To achieve this, we replace the sensitivity field along each ray with its average value as illustrated in Figure 8.22. If this average value is above cutoff, then all material along this ray is retained, if not they are removed.

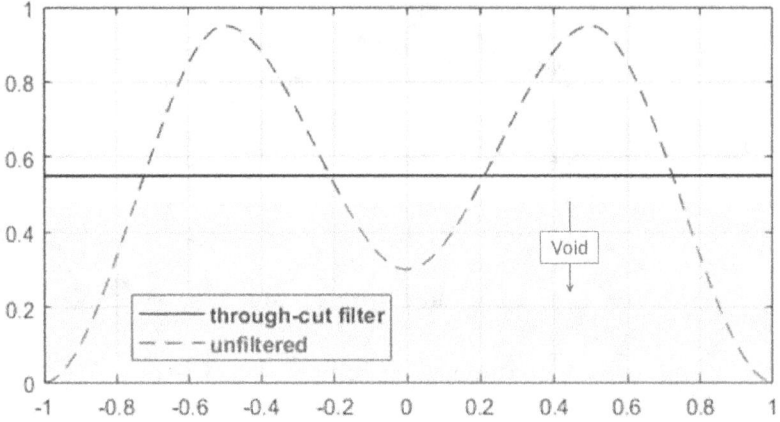

Figure 8.22: Effect of through-cut filter on sensitivity field.

Figure 8.23 demonstrates the effectiveness of the through-cut constraint.

Figure 8.23: Optimized design with through-cut constraint.

■ **Example 8.3 Through-Cut constraint for the swing arm**

1. Load the *SwingArmProject.prj*.
2. Under *TopOpt Constraint* menu, set *Through Cut* to *ZDir*.
3. Optimize to find the lowest possible volume considering constraints on displacement and stress.

Figure 8.24: Final topology with through cut constraint.

Ⓡ Note that the cross-section of the optimized design remains constant

8.3 Retaining Surfaces

in Z direction.

(R) The final volume fraction is 0.7, where safety factors on displacement and stress are 1.47 and 1.61, respectively. This means that considering these constraints no further meaningful improvement could be made on the design.

8.3 Retaining Surfaces

Often, there may be surfaces on the initial design that must be retained for manufacturability, accessibility, ease of assembly, safety, or aesthetics. Implementation of this constraint is straightforward, in that the sensitivity field at these surfaces must be made larger than the cut-off. This is illustrated in Figure 8.25.

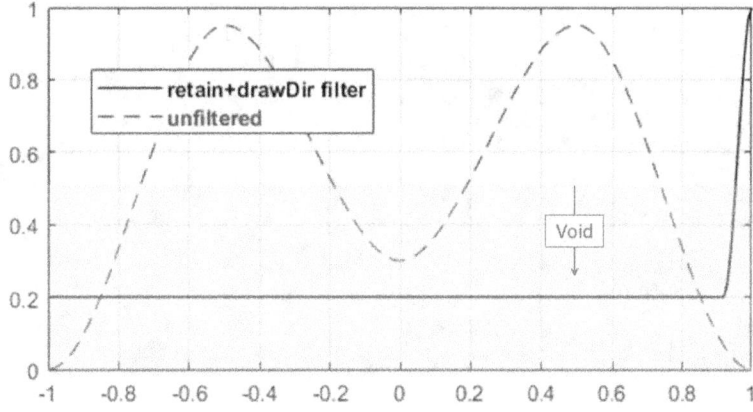

Figure 8.25: Effect of surface retain constraint on sensitivity field.

■ **Example 8.4 Retain constraints for the swing arm**

1. To impose such constraints, switch to geometry mode.
2. Make sure the symmetry along Z is turned-off.
3. Next, set *Selection* to *Plane* and select the three surfaces shown in Figure 8.26a.
4. Scroll down and select the *Retain* option under *Load Type* (see Figure 8.26b), and apply.

(R) This constraint states that, during optimization, material should not be removed near these surfaces.

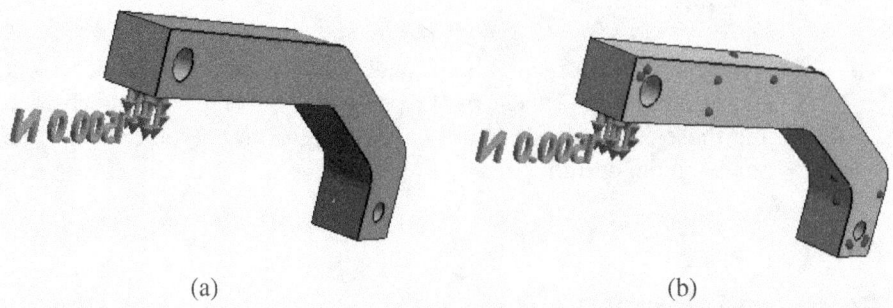

Figure 8.26: (a) Selecting the three faces, (b) resulting plot.

5. Next, impose the draw-direction constraint along Z.
6. Now, set the desired volume fraction to 0.1 and optimize. The resulting topology is illustrated in Figure 8.27.

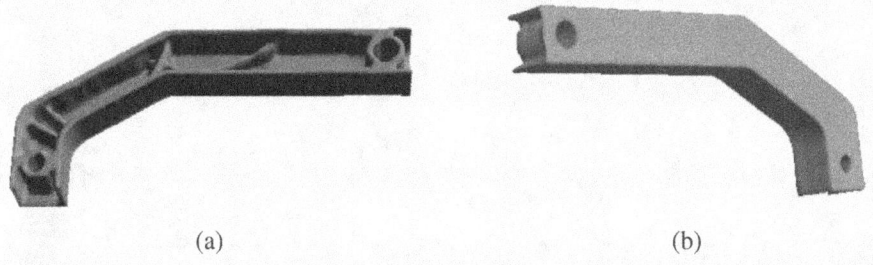

Figure 8.27: Final topology, with retained surfaces.

(R) Observe that the surfaces have been retained, and material has been carved out from the inside.

7. The maximum vonMises stress is 364 MPa and the safety factor is only 1.37.

Figure 8.28: Stress distribution on optimized design.

8.3 Retaining Surfaces

■ **Example 8.5 Knuckle Project**

1. Set units to MKS.
2. Load the *Knuckle.stl*.
3. Apply alloy steel as material.
4. Fix the two bottom circular holes.
5. Apply a torque of $100N.m$ on the outer top circular face.

Figure 8.29: Loading condition of the Knuckle subject to torsion.

6. Set mesh quality to medium.
7. Impose symmetry along X and Z and solve the FEA.

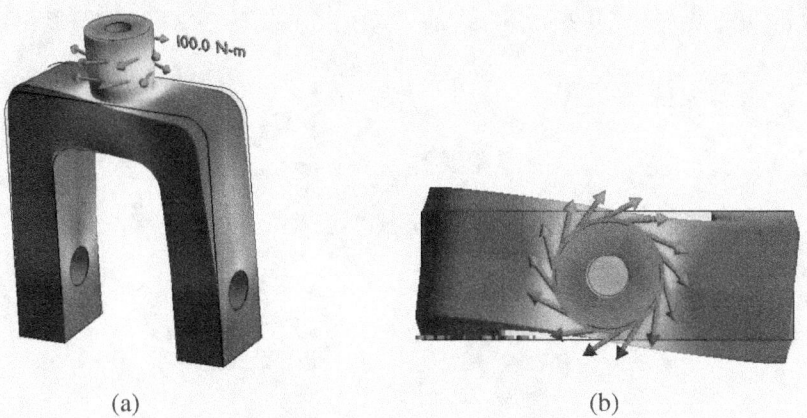

(a) (b)

Figure 8.30: Knuckle FEA solution.

8. Make sure the fixed faces are retained.
9. Optimize the design to 0.3 volume fraction. The maximum vonMises stress is 82.7 MPa and the safety factor is about 6.

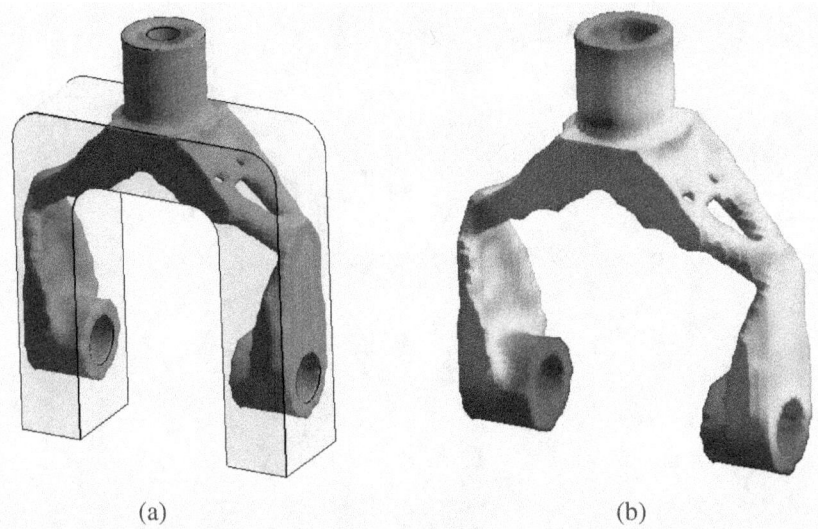

Figure 8.31: (a) Optimized knuckle at 0.3 volume fraction, (b) Stress distribution of the optimized knuckle.

10. Save the project as *KnuckleProject*.
11. Next, we will impose retain constraint. To this end, change the *Selection* Option under *Loads* menu to *Plane*

    ® This will allow you to set select triangles that lie on a plane.

12. Select the surface highlighted in Figure 8.32.
13. Apply the *Retain* condition. The face will be retained during optimization.

Figure 8.32: Retain face selection for the Knuckle example.

14. Now optimize with the project parameters.
15. The final optimized design (0.3 volume fraction) is illustrated in Figure 8.33.

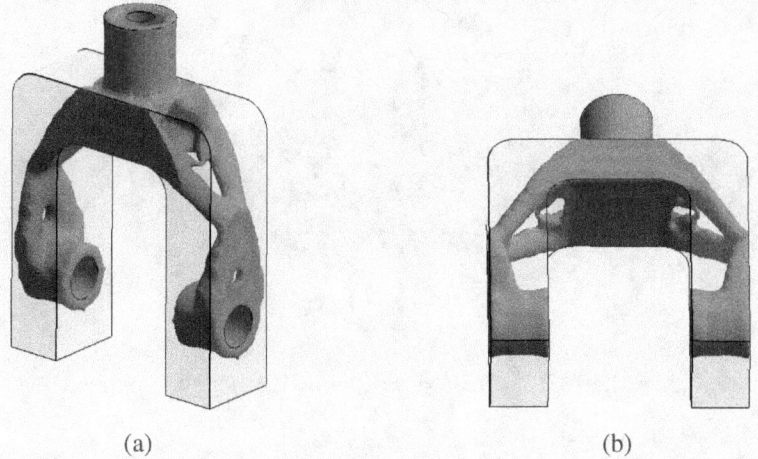

(a)　　　　　　　　　　　　(b)

Figure 8.33: Optimized Knuckle under torsion at 0.3 volume fraction with retained surface.

16. The maximum vonMises stress is 76 MPa and the safety factor is about 6.57.

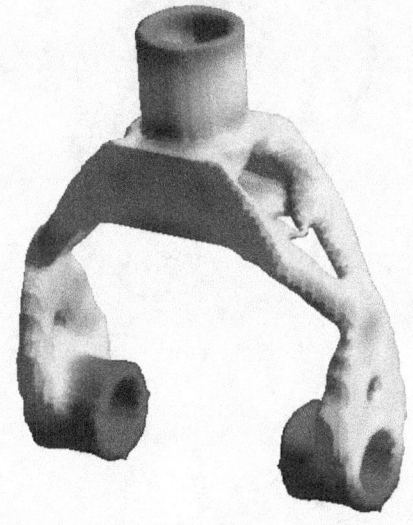

Figure 8.34: Stress distribution of optimized knuckle.

(R) Observe that the selected face has been retained.

8.4 Cyclic Symmetry

In this section we explain the cyclic constraint. The basic idea is similar to previously discussed constraints, that is manipulate the sensitivity field to meet the desired constraint.For instance, consider the following optimized disk, where we have imposed circular symmetry if 120°. In other words, we want the final design to comprise of three similar sectors.

Figure 8.35: Optimized design with 120° cyclic symmetry.

To examine the change in sensitivity, let us consider the following circular cross-section:

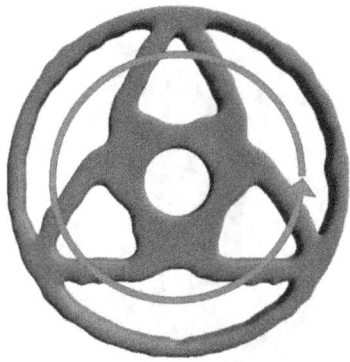

Figure 8.36: Consider the circular cross-section.

Along this cross-section, the topological sensitivity is illustrated below:

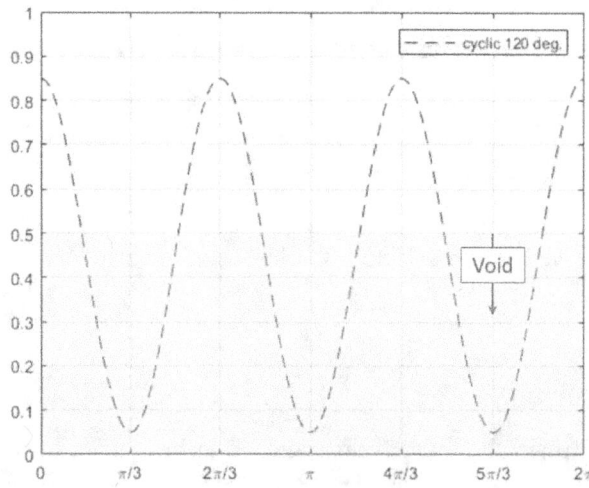

Figure 8.37: Effect of cyclic constraint on sensitivity field with 120° cyclic symmetry.

8.4 Cyclic Symmetry

Thus, for every point of design, we find the two other points that are 120 degrees apart, along the same radius, and assign identical (average) sensitivity values. One can also impose a cyclic symmetry constraint of 60°, which would create the following optimized design below.

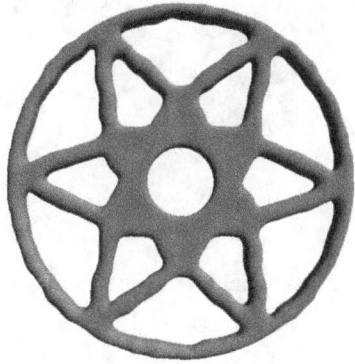

Figure 8.38: Optimized design with 600° cyclic symmetry.

Figure 8.39 illustrates how cyclic symmetry constraint alters the topological sensitivity field for two cases of 120° and 60°.

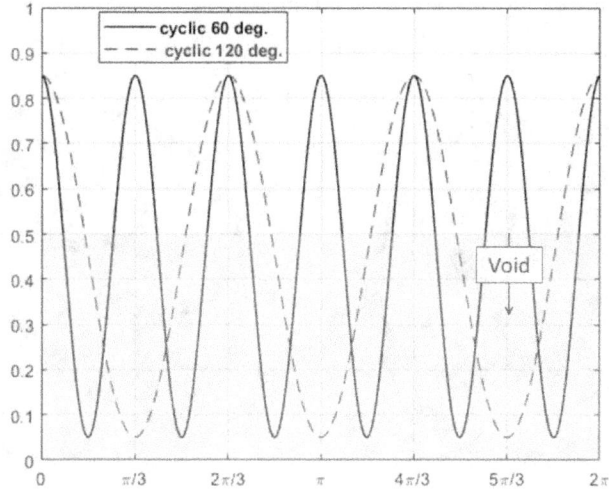

Figure 8.39: Effect of cyclic constraint on sensitivity field.

■ **Example 8.6 Cyclic Constraints for the wheel**

1. Load the *CircularPlateHole.prj*. As shown in Figure 8.40, inner circular face is fixed and a torsional load of 200 lb-in is applied on the outer face.

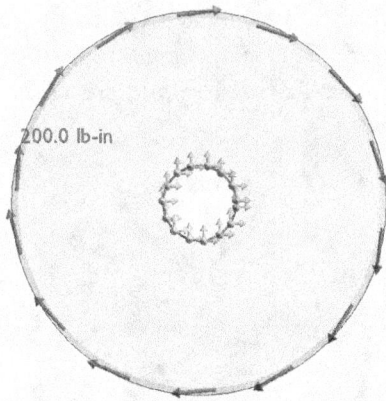

Figure 8.40: Circular wheel geometry and loading condition.

2. Under *TopOpt Constraint*, set *Cyclic Sym (Z)* to *3* (120 degrees).

   (R) By imposing this constraint, we are essentially dividing the circular plate to 3 identical sectors with central angles of 120 degrees.

3. Optimize with the default project settings, Figure 8.41a shows the optimized design with 120 degrees cyclic symmetry.
4. Now, change *Cyclic Sym (Z)* to *6* (60 degrees) and optimize. Figure 8.41b shows the optimized design with 60 degrees cyclic symmetry.

(a)   (b)

Figure 8.41: Final topology with (a) 120 degrees and (b) 60 degrees cyclic constraint.

8.5 Minimum Feature Size

The last manufacturing constraint we consider is *minimum feature size*, which controls the size of smallest feature within the optimized design. This constraint is directly related to manufacturing process and its capabilities in fabricating thin features. Same as before, to impose minimum feature size constraint, we must modify the sensitivity field. In this case,

8.5 Minimum Feature Size

sensitivity value at each point (or node) is replaced by its neighborhood average, where the radius r of the neighborhood controls the feature size. Obviously, as the radius increases, fine details are eliminated and feature size increases. Figure 8.42 illustrates the basic idea behind this widely used filtering scheme.

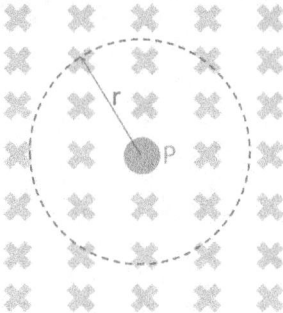

Figure 8.42: Sensitivity smoothing filter.

Consider the following optimized Messerschmidt-Bölkow-Blohm (MBB) beam, where thin features with maximum size of 5cm are allowed.

Figure 8.43: Unfiltered optimized design.

Now, let us consider the central section of the optimized designs; as shown in Figure 8.44.

Figure 8.44: Consider the central section of the optimized design.

Figure 8.45 illustrates the effect of smoothing filtering on feature sizes to ensure manufacturability of the design:

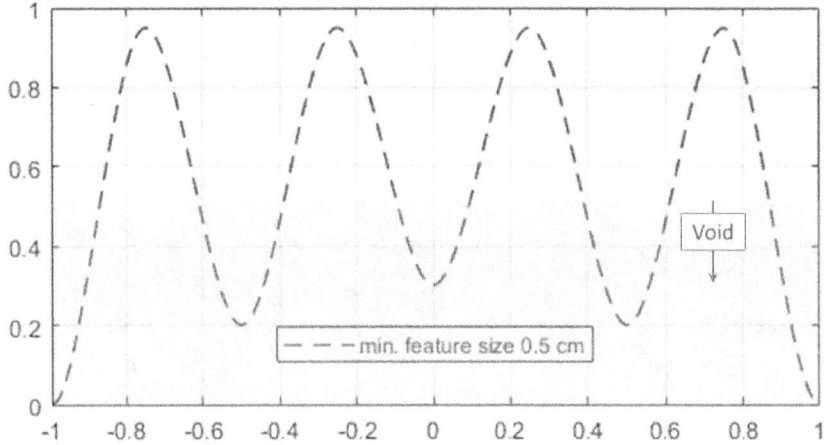

Figure 8.45: Effect of minimum feature size constraint on sensitivity field.

Next, let us consider the case where the manufacturing process is not as accurate and the maximum resolution is 10 cm. Therefore, we would need to alter the design to something similar to Figure 8.46:

Figure 8.46: Unfiltered optimized design.

Considering the same central section as before:

Figure 8.47: Consider the central section of the optimized design.

Figure 8.45 illustrates the effect of smoothing filtering for these two cases where as filter radius increases, the number of features are reduced while the thickness of each feature increases.

8.5 Minimum Feature Size

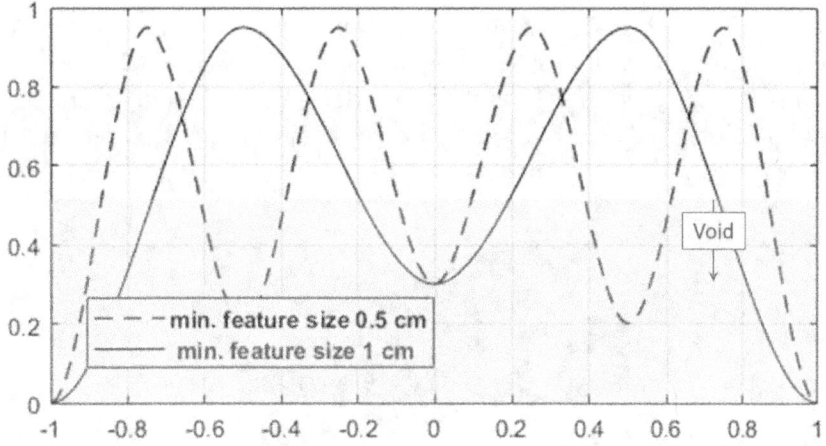

Figure 8.48: Effect of minimum feature size constraint on sensitivity field.

■ **Example 8.7** **Minimum feature size for the Messerschmidt-Bölkow-Blohm (MBB) beam**

1. Load the *MBB.prj*.
2. Set mesh quality to *Medium*.
3. Under *TopOpt Constraints*, change *RelMinFeatSize* to 2.
4. Optimize to 0.6 volume fraction.
5. Next, change *RelMinFeatSize* to 3, and optimize to the same 0.6 volume fraction.

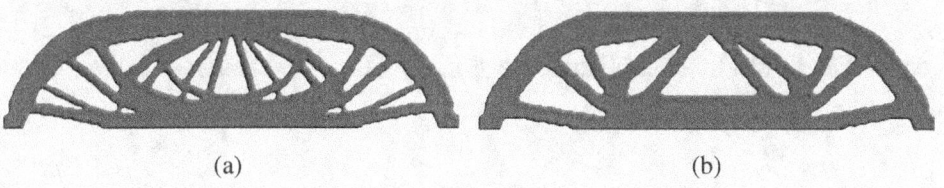

(a)   (b)

Figure 8.49: Optimized MBB beam at 0.6 volume fraction with
(a) thin features, and (b) thick features.

8.6 Exercises

Exercise 8.1 Though this exercise, you will study that multiple design and manufacturing constraints can be imposed, provided they do not conflict with each other. Load the project *ThickPlateProject.prj*. Retain the two surfaces marked on in Figure 8.50 (using boundary conditions) and apply draw direction constraint along z. Optimize and compare the results aginst those obtained previously

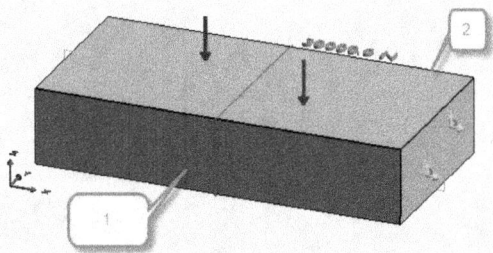

Figure 8.50: Thick plate project with surfaces to be retained.

Exercise 8.2 Though this exercise, you will study that performance and manufacturing constraints can be imposed simultaneously. Load the project *SwingArmProject.prj*. Check that the first eigenvalue is around 515 Hz. Impose an eigenvalue constraint of 500 Hz, impose a draw direction constraint along z, and optimize for a volume fraction of 0.1; compare the results obtained previously. Does changing the objective affect the final design?

Exercise 8.3 Consider the spindle mount example discussed in the previous chapter; retain surface-1 identified in Figure 8.51. Carry out the optimization as before. Compare the topologies, and safety factors.

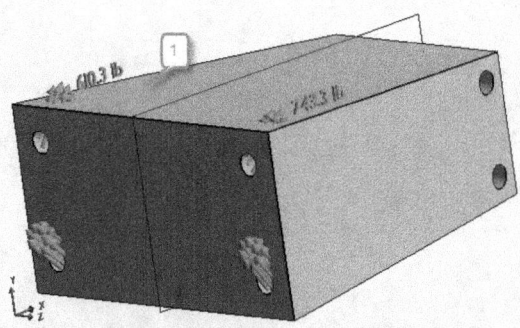

Figure 8.51: Spindle mount project with surface to be retained.

8.6 Exercises

Exercise 8.4 Consider again the spindle mount example. This time, instead of retaining the surface, impose draw direction along y. Carry out the optimization as before, and compare the results.

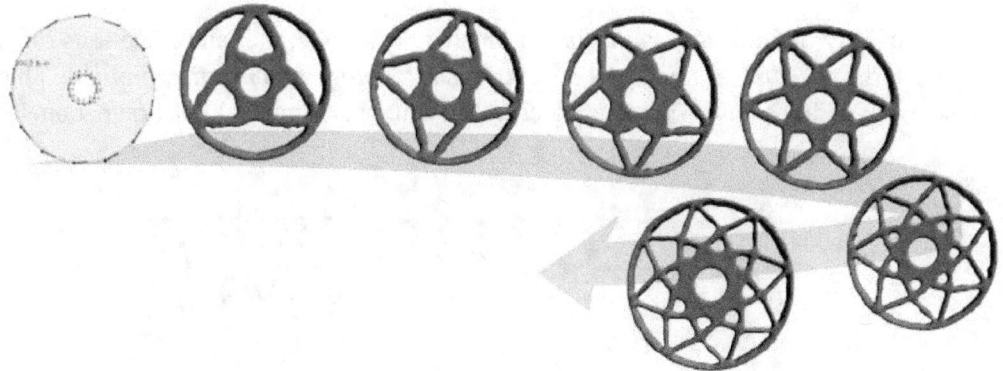

9. Non-Uniqueness of Designs

The main objective of this chapter is to emphasize that there is no unique solution in topology optimization, i.e., for the same set of loads, there may be distinct, and equally valid topologies, that satisfy all the constraints. This is a well-known fact in engineering design. For example, consider the GE-GrabCAD design problem that was posed several years ago; see Figure 9.1.

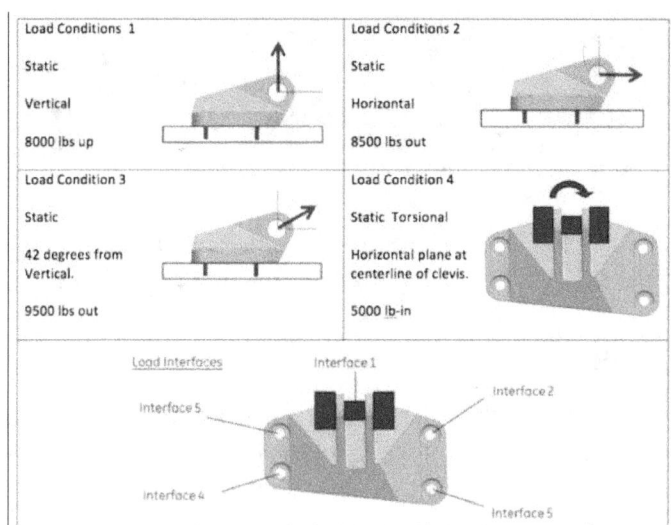

Figure 9.1: GE-GrabCAD design problem (https://grabcad.com/challenges/ge-jet-engine-bracket-challenge).

The above problem attracted hundreds of designers from around the world. A variety of designs were submitted (see Figure 9.2), and many of these distinct designs were equally good.

Topology optimization is just an automated design method, and it exhibits the same characteristic: *there are numerous solutions to a topology optimization problem.* One can discover several *equally good* topologies by varying one or more simulation parameters.

Figure 9.2: Variety of designs submitted
(https://grabcad.com/challenges/ge-jet-engine-bracket-challenge).

9.1 Mesh Density

First, we will consider varying the mesh density. As a general rule, increasing the number of elements will: (1) increase the complexity of the design, (2) increase the computational time, and (3) decrease the safety factors since the stresses will be evaluated at a finer resolution.

■ **Example 9.1 Mesh density, a comparison study**

1. First load the project *LBracketProject.prj*.
2. With the default setting of 10,000 elements, optimize the design for 0.5 volume fraction.

9.2 Minimum Feature Size

3. Then, repeat the optimization for 25,000, 50,000, 150,000, 250,000 and 500,000 elements.
4. Table 9.1 summarizes the results. For 250K elements, the minimum feature size was increased to 3. Observe that despite the differences in these designs, they exhibit similar performance; for example, the stress safety factor remains between 2.40 and 2.75.

Table 9.1: Mesh Density

Optimized	Mesh Density	Displacement Safety Factor	Stress Safety Factor	Time (secs)
	10K	14.4	2.86	14
	25K	15.05	2.5	33
	50K	15.3	2.23	65
	150K	15.3	2.16	237
	500K	15.5	2.09	920

9.2 Minimum Feature Size

Next, we will consider the impact of minimum feature size on the optimized topology. In general, increasing the feature size will decrease the complexity of the topology.

Example 9.2 Minimum feature size, a comparison study

1. First load the project *LBracketProject.prj*.
2. Set the number of elements to 150,000, and the relative feature size to 1. Optimize the design for 0.5 volume fraction.
3. Then, repeat the optimization for relative feature size of 2, 3, 4 and 5.

Table 9.2: Minimum Feature Size

Optimized	Feature Size	Displacement Safety Factor	Stress Safety Factor
	1	15.4	2.2
	2	15.3	2.2
	3	15.3	2.12
	4	15.2	2.12
	5	15.1	2.10

9.3 Step Size

Next, we will consider the impact of step-size used by the Pareto algorithm on the final topology. The step-size is the maximum volume fraction decrement that the Pareto algorithm uses to trace the Pareto curve. As expected, smaller the decrement, larger the computational

9.4 Cyclic Symmetry

time. Further, more intermediate topologies are saved. It may also be possible to discover new topologies by changing this parameter.

■ **Example 9.3 Optimization step size**

1. Load the project *SwingArmProject.prj* saved previously.
2. Impose draw direction along z.
3. Optimize.
4. Next change the step-size to 0.01, and optimize
5. Finally change the step-size to 0.05, and optimize
6. Table 9.3 summarizes the results. Smaller step-size typically leads to finer features.

Table 9.3: Step Size

Optimized	Step Size	Displacement Safety Factor	Stress Safety Factor
	0.01	1.09	1.54
	0.025	1.01	1.48
	0.05	1.0	1.37

9.4 Cyclic Symmetry

Finally, we will consider the impact of cyclic symmetry on the topology.

■ **Example 9.4 Cyclic symmetry, a comparison study**

1. First reload the project *CircularPlateHole.prj* saved previously (see Figure 8.40).
2. Optimize the design for 0.5 volume fraction and different symmetry angles of 120 (3 sectors), 90, 72, 60, and 45 degrees.
3. Table 9.4 summarizes the results. Observe that despite the apparent geometric differences in these designs, they exhibit similar performance; where stress safety factor remains almost a constant.

Table 9.4: Cyclic Symmetry

Optimized	Symmetry Angle	Displacement Safety Factor ($\times 100$)	Stress Safety Factor ($\times 100$)
	120	13.8	1.5
	90	18.2	1.4
	72	19.9	1.4
	60	19.2	1.5
	45	21.1	1.5

9.5 Exercises

Exercise 9.1 Load the project *CircularPlateHoleProject.prj*. For a fixed cyclic symmetry of 5, vary the step size, optimize and compare the results. Similarly, for a fixed symmetry, fixed step size, vary the mesh size, optimize and compare the results.

Exercise 9.2 Load the project *KnuckleProject.prj*. Vary the step size, optimize and compare the results. Next vary the mesh size, optimize and compare the results.

9.5 Exercises

Exercise 9.3 Load SpindleMountProject.prj; study the topologies obtained by varying the number of mesh elements.

Exercise 9.4 Load SpindleMountProject.prj; this time, study the effect of the step size on the topology.

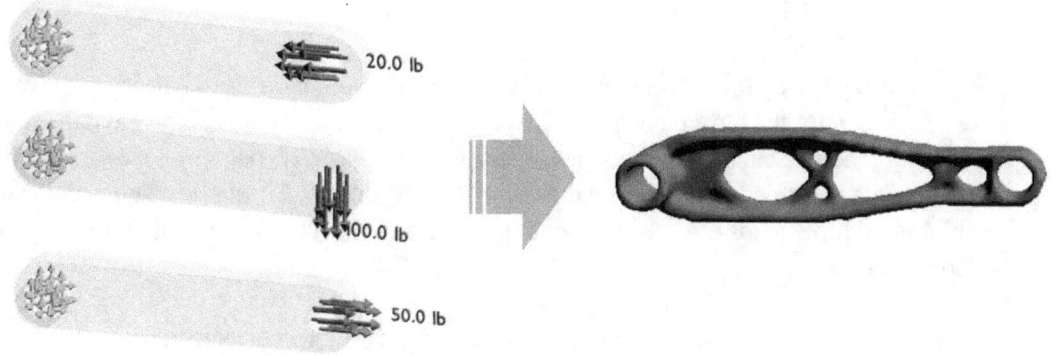

10. Multi-Load Optimization

Most industrial applications require designs to retain their structural integrity under multiple load cases. In this chapter, we will discuss the topology optimization of such multi-load problems. A typical example of a multi-load problem is the bi-cycle crank, illustrated in Figure 10.1. During a full pedaling cycle, the crank arm passes through four distinct positions. At each position, it experiences a different loading condition. At position A where the pedal is passing through the highest point, it sees a compressive load. At position B where the crank arm is horizontal, it sees a bending load. At position C, it sees a tension force. At position D, the load is negligible. The objective of multi-load topology optimization is to arrive at a single light-weight crank arm design that can handle all the loads.

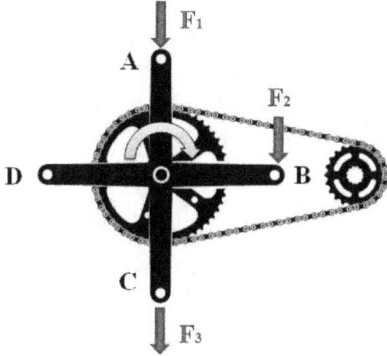

Figure 10.1: The crank arm subject to multi-load during a pedaling cycle; (ref: Bini. et. al, Journal of Science and Cycling, vol 2., 2013).

10.1 Multi-Load Strategies

One possible approach for solving multi-load optimization problems is to consider the worst case, i.e., design for the most demanding load case. For example, one can argue that the most demanding loading scenario in the above example is Scenario B (bending load). However, this strategy is incorrect, i.e., it can lead to an erroneous design. Similarly, one can consider combining all three loads into a single load. Once again, it is easy to show that this method is also flawed.

Instead, one must dynamically weigh the sensitivity fields to correctly arrive at the optimized design. As an illustration, consider the 2D multi-load problem in Figure 10.2. Let us suppose the objective is to minimize the volume of the design, subject to compliance constraints for the two loads.

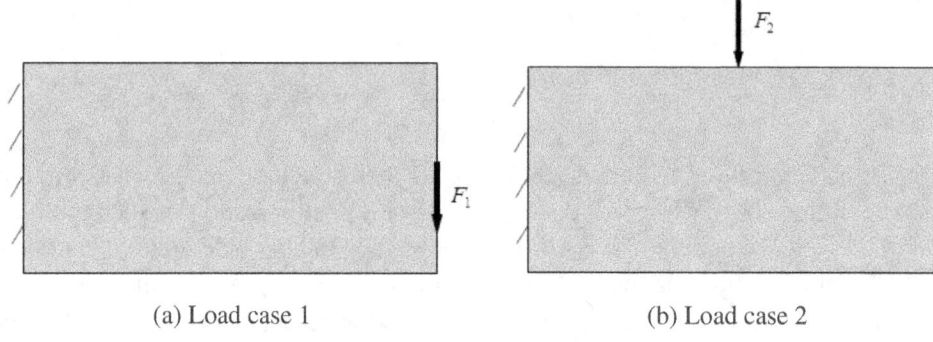

(a) Load case 1 (b) Load case 2

Figure 10.2: A multi-load optimization problem.

Mathematically, the above problem can be posed as:

$$\begin{array}{c} \underset{\Omega \subset D}{\text{minimize}} \quad V \\ J_1 \leq J_1^{max} \\ J_2 \leq J_2^{max} \end{array} \qquad (10.1)$$

The first step is to compute the compliance topological sensitivity for the two loads. Recall that the closed-form expression for compliance topological sensitivity is given by:

$$\mathcal{T}(p) = \frac{4}{1+v} \sigma(\mathbf{u}) : \varepsilon(\lambda) - \frac{1-3v}{1-v^2} tr(\sigma(\mathbf{u})) tr(\varepsilon(\lambda)) \qquad (10.2)$$

This two sensitivity fields are illustrated in Figure 10.3.

10.1 Multi-Load Strategies

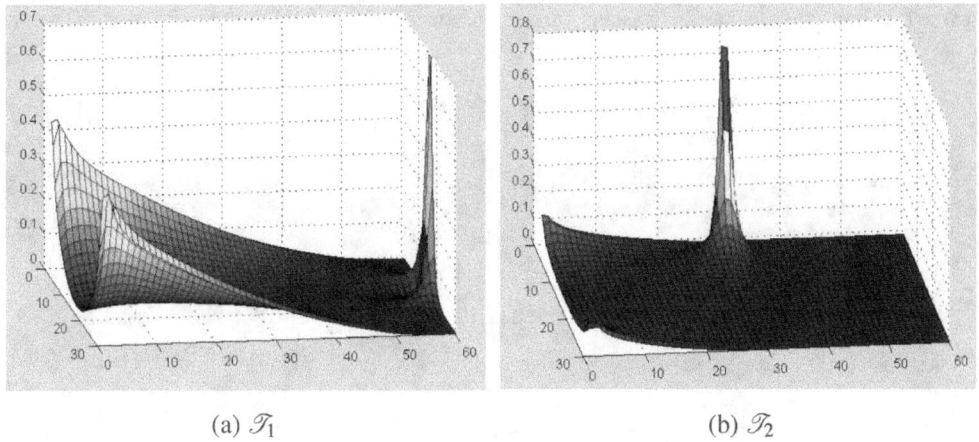

Figure 10.3: Topological sensitivity fields for the two load cases.

Using this expression, one can compute the two topological sensitivity fields $\mathcal{T}_1$ and $\mathcal{T}_2$. To optimize the design, we must account for both sensitivity fields. If equal emphasis is placed on both loads, one can define a new sensitivity field as the average of the two, i.e.:

$$\mathcal{T}_{avg} = \frac{\mathcal{T}_1 + \mathcal{T}_1}{2} \tag{10.3}$$

The resulting field is illustrated in Figure 10.4.

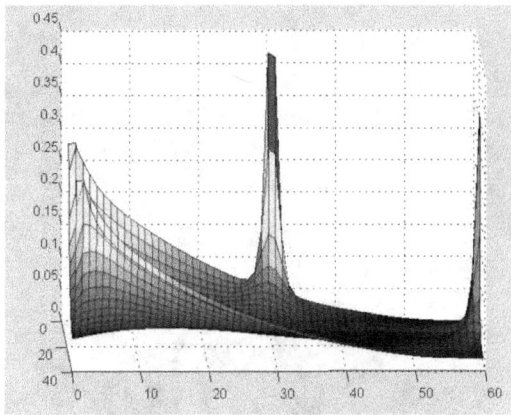

Figure 10.4: The average topological sensitivity fields.

However, the simple averaging does not take into account the constraints in Equation 10.1. Observe that, in the above example, if the constraint on J_1 is more restrictive than that on J_2, one should weigh $\mathcal{T}_1$ more than $\mathcal{T}_2$. In other words, material removal should be dynamically guided. Thus, suppose the current compliances for the two structural problems are J_1 and J_2, then the two weights may be defined in a generic fashion as:

$$\begin{aligned} w_1 &= g(J_1/J_1^{max}) \\ w_2 &= g(J_2/J_2^{max}) \end{aligned} \tag{10.4}$$

where g(.) is a positive non-decreasing function. The weighted sensitivity field is then defined as:

$$\mathcal{T} = w_1 \mathcal{T}_1 + w_2 \mathcal{T}_2 \tag{10.5}$$

For example, a linear-weighting function $g(x) = x$ leads to:

$$\mathcal{T} = (J_1/J_1^{max}) \mathcal{T}_1 + (J_2/J_2^{max}) \mathcal{T}_2 \tag{10.6}$$

A non-uniform weighting may result in a field illustrated in Figure 10.5:

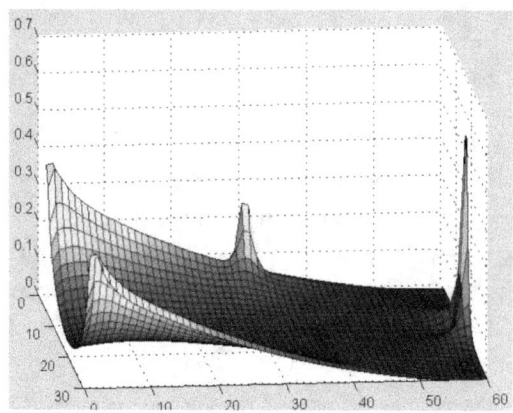

Figure 10.5: The linearly-weighted topological sensitivity field.

10.2 Design of a Crank Arm

■ **Example 10.1 Imposing multiple load conditions on the Crank Arm Model**

1. Change the units to IPS, since the crank model (.stl) file was saved in inches.
2. Load the *CrankArm.stl* file from *ParetoWinExamples* folder; the model will be loaded as in Figure 10.6.

Figure 10.6: *CrankArm* model.

3. Change the material to *Al 1060*, increase the allowable stress to 30 ksi, and apply.
4. Next, we will apply restraints on the model. Select the cylindrical hole as shown in Figure 10.7a.
5. Then, as in earlier chapters, under the *Structural Loads* menu, set the *Selection* to *XYZFixed* and Apply. The face is restrained in all three directions, as in Figure 10.7b.

10.2 Design of a Crank Arm

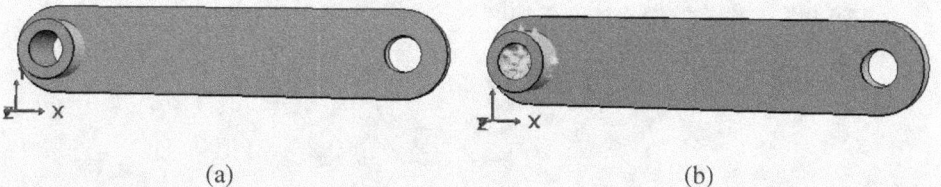

(a)   (b)

Figure 10.7: (a) Face selection, and (b) restraint condition.

6. Next, select the face as illustrated in Figure 10.8a.
7. Then, apply a compressive force of 20 lbs in the negative x-direction as in Figure 10.8b; the resulting plot is illustrated in Figure 10.8c.

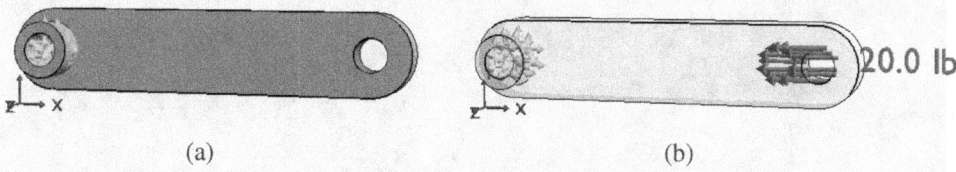

(a)   (b)

Figure 10.8: (a) Face selection, (b) applied loading conditions.

We will now increment the Load-set to 1 as in Figure 10.9a.

> (R) Observe that the restraint is carried over, but not the force, i.e., all load sets share the same set of restraints.

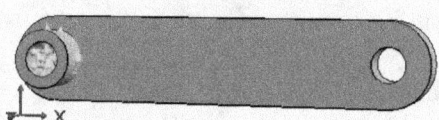

Figure 10.9: The restraint is carried over.

8. For this load set, select the face again as illustrated in Figure 10.10a.
9. Then, apply a force of 100 lb in the negative y-direction as in Figure 10.10b and Figure 10.10c.

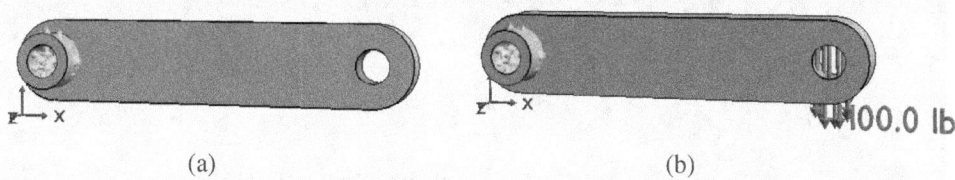

(a)   (b)

Figure 10.10: (a) Face selection, and (b) applied loading conditions.

10. Next, increment the Load-set to 2 as in Figure 10.11a. Once again, observe that the

restraint is carried over, but not the force.

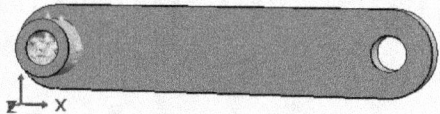

Figure 10.11: The restraint is carried over.

11. For the final load set, select the face again as in Figure 10.12a.
12. Then, apply a tensile force of 50 lb in the x-direction as in Figure 10.12b and Figure 10.12c.

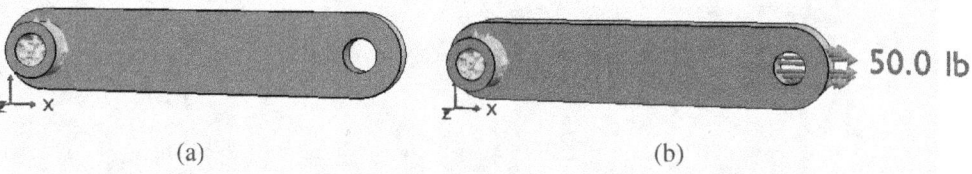

(a)   (b)

Figure 10.12: (a) Face selection, and (c)applied loading conditions.

10.3   FEA on a Multi-Load Problem

Now that we have applied all the three loads, we will move on to finite element analysis. We will carry out three different FEA, one for each load.

■ **Example 10.2 FEA on the multi-load Crank Arm Model**

1. Set the Load Set to 0.
2. Carry out an FEA with a medium mesh quality. The results are summarized in Figure 10.13b and Figure 10.13c.

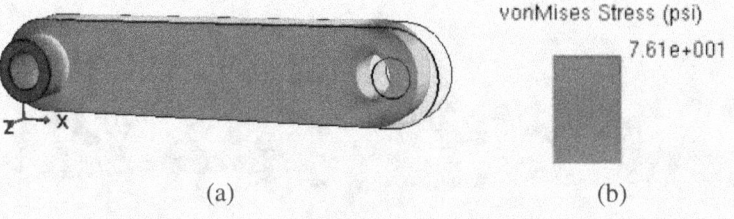

(a)   (b)

Figure 10.13: (a) stress results, and (b) max stress.

3. Set the Load Set to 1, and repeat FEA; see Figure 10.14 for results.

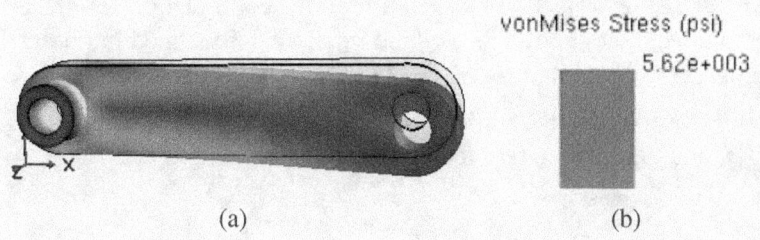

Figure 10.14: (a) Stress results, and (b) max stress.

4. Set the Load Set to 2, and repeat FEA; see Figure 10.15 for the results.

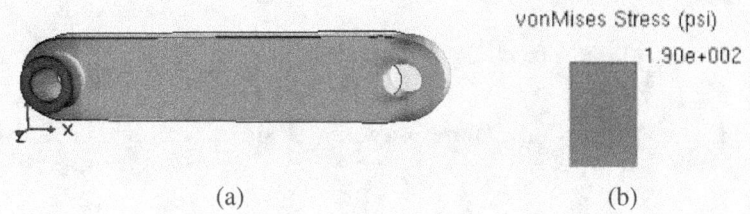

Figure 10.15: (a) Stress results, and (b) max stress.

(R) As one can expect, the stress is the highest for the bending load.

5. You can now save the work the material, i.e. restraints and loads). When prompted, save the project as *CrankArm.prj*.

10.4 Optimizing a Multi-Load Problem

We will now carry out a multi-load topology optimization, where the optimizer simultaneously considers all of the prescribed loading conditions to ensure that all of the constraints are satisfied.

■ **Example 10.3 Optimizing the multi-load Crank Arm Model**

1. Under *TopOpt Constraints* menu:
   - The *Draw Direction* is set to Z, i.e., the design can be cast in the z-direction.
   - The *Keep Fixed Faces* option means that material surrounding the two restrained holes will be retained during optimization,
2. Under *Optimize Topology* menu, set the desired volume fraction is set to 0.1.

   (R) Multi-load optimization is computationally very demanding since multiple FEAs must be carried out for each load set.

3. Before optimizing, you should save the project to store all the optimization parameters.

4. After saving the project, you can carry out the optimization.
5. The optimized model of 0.31 volume fraction is illustrated in Figure 10.16a, and the final safety factors are illustrated in Figure 10.16b.

   ® Observe that the critical load is Load Set #1, i.e., the bending load, and the stress safety factor at termination is 1.45.

Vol. Frac: 0.26
Time Taken: 110.6 (secs)
Disp. Safety Factor = 2.98
Critical load = 1
Stress Safety Factor = 1.12
Critical load = 1

(a) (b)

Figure 10.16: (a) optimized design, (b) critical safety factors and critical load.

6. One can also recover other Pareto-optimal designs, with higher safety factors as illustrated in Figure 10.17.

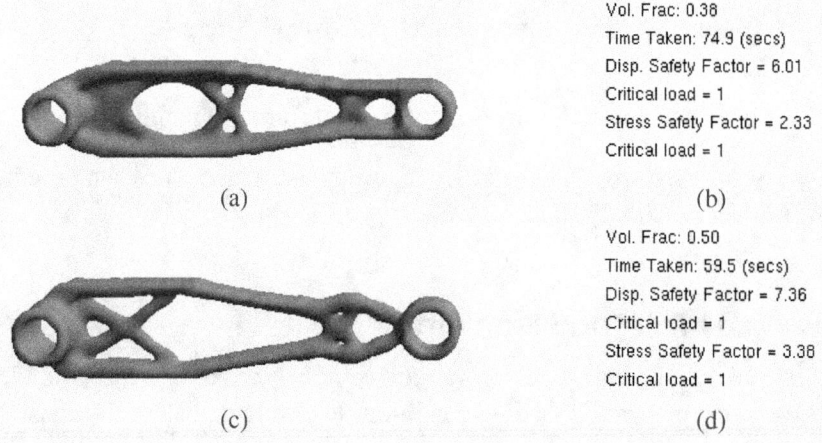

Vol. Frac: 0.38
Time Taken: 74.9 (secs)
Disp. Safety Factor = 6.01
Critical load = 1
Stress Safety Factor = 2.33
Critical load = 1

(a) (b)

Vol. Frac: 0.50
Time Taken: 59.5 (secs)
Disp. Safety Factor = 7.36
Critical load = 1
Stress Safety Factor = 3.38
Critical load = 1

(c) (d)

Figure 10.17: Other Pareto-optimal designs.

10.5 Post-Optimize FEA

For multi-load problems, after optimization, we will carry out a number of FEAs, one for each load.

■ **Example 10.4 Optimizing the multi-load Crank Arm Model**

1. Set the Load Set 0,
2. Unselect *Remesh* and carry out an FEA as in Figure 118a. The results are summa-

rized in Figure 10.18. The maximum stress is 0.85 ksi.

® If the *Remesh* option is unselected, then FEA is carried out on the optimized mesh.

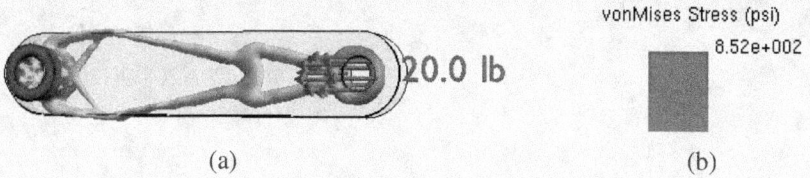

(a) (b)

Figure 10.18: Re-mesh and solve the optimized design with load set 0.

3. Set the Load Set to #1, and repeat FEA; see Figure 10.19 for the results. The maximum stress is 27 ksi.

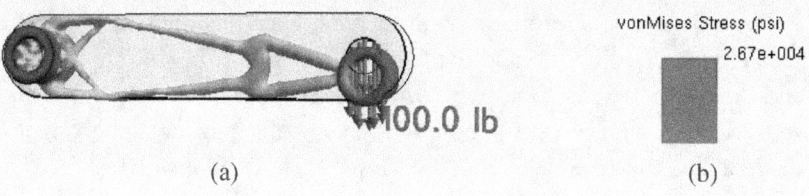

(a) (b)

Figure 10.19: Re-mesh and solve the optimized design with load set 1.

4. Set the Load Set to #2, and repeat FEA; see Figure 10.20 for the results. The maximum stress is 2.1 ksi. As expected, the dominating load is the bending load (Load Set #1).

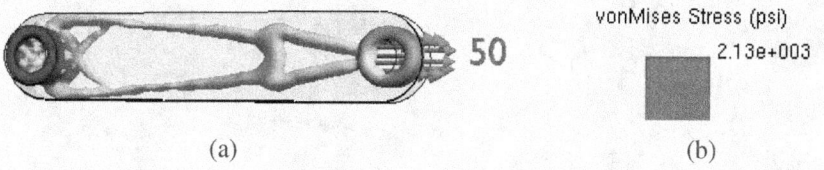

(a) (b)

Figure 10.20: Re-mesh and solve the optimized design with load set 2.

10.6 Exercises

Exercise 10.1 Pose the two-load problem on the LBracket as illustrated in Figure 10.21; the material is Alloy Steel. Using a medium-quality mesh, optimize the design with respect to both loads. How does the design change if the first force is doubled? Repeat this exercise when the second force is doubled instead.

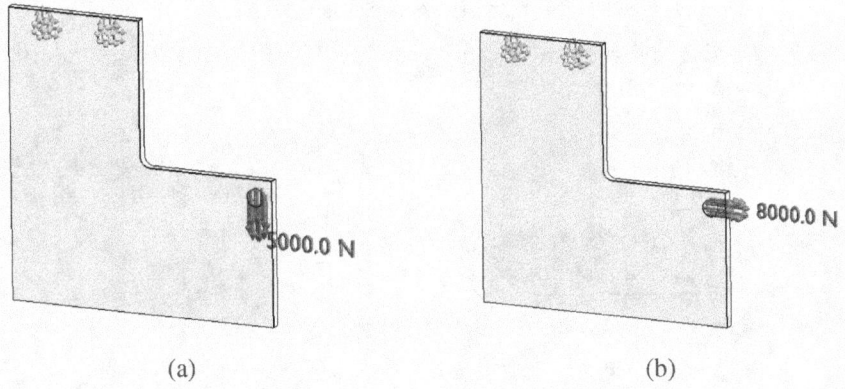

Figure 10.21: Multi-load problem over the L-Bracket geometry.

Exercise 10.2 Pose the two-load problem on the Knuckle as illustrated in Figure 10.22 where load-0 is a torque of 100 N-m, and load-1 is a force of 13,000 N. The material is Alloy Steel, and the part is fixed at the two horizontal holes. Using a medium-quality mesh, optimize the design with respect to both loads. How does the design change if the force is doubled? Repeat this exercise when the torque is doubled instead.

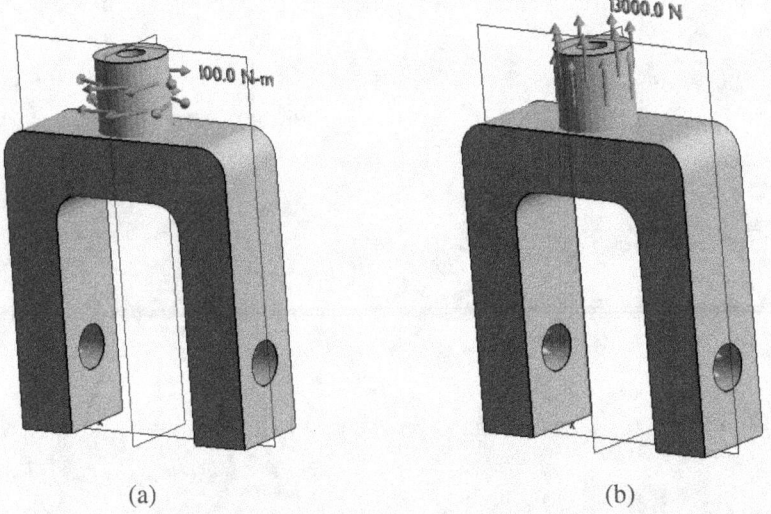

Figure 10.22: Multi-load problem over the knuckle geometry.

Exercise 10.3 Consider again the spindle mount example. Pose the two-load problem illustrated in Figure 10.23. All other problem parameters being the same. Optimize the design.

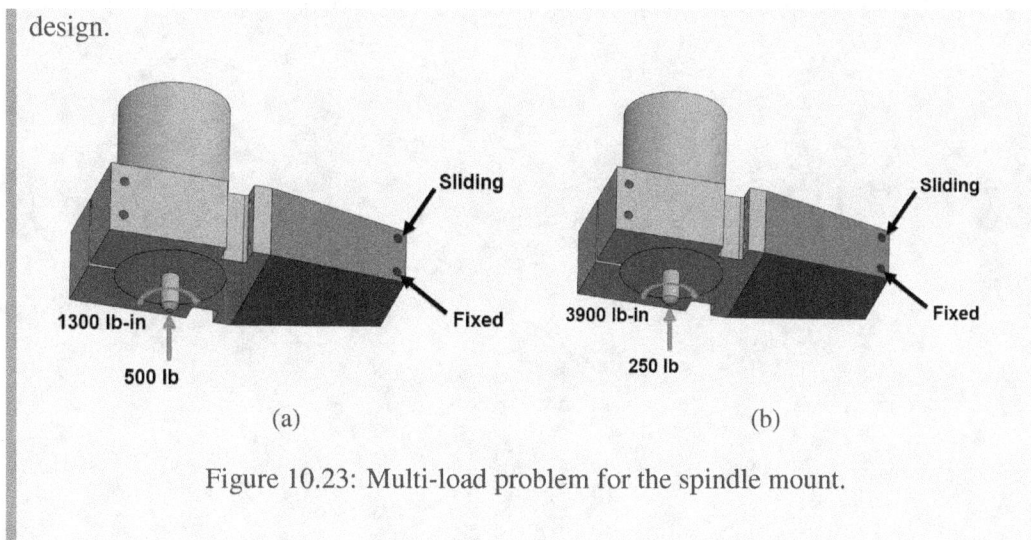

Figure 10.23: Multi-load problem for the spindle mount.

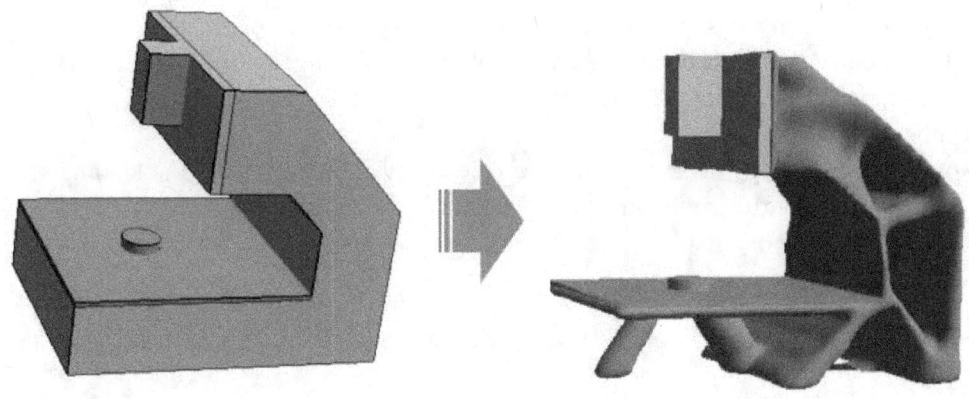

11. Multi-Body & Multi-Material TO

In this chapter, we will learn how to pose multi-body and multi-material topology optimization problems.

11.1 Multi-Body Optimization

In a multi-body problem, there are usually several parts, and only a subset of the parts must be optimized. For example, consider the spindle mount problem posed earlier; see Figure 11.1. Recall that, although the assembly consisted of three parts, we were only interested in optimizing the spindle mount. Previously, we isolated the spindle mount from the assembly, transferred the moments and torques to pose a single-body problem. This is certainly an acceptable strategy. However, it is more convenient to directly consider the assembly, and let the finite element model handle the transfer of forces, etc. We will illustrate below how this can be carried out.

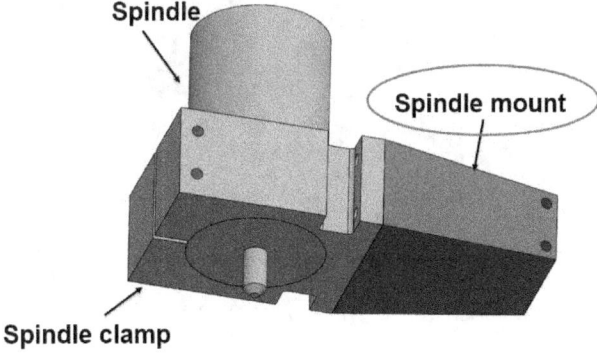

Figure 11.1: A multi-body optimization problem where only the spindle-mount must be optimized.

From an implementation perspective, multi-body optimization is straightforward: once a finite element analysis has been carried out, the topological sensitivity field is computed only in the part that must be optimized. For all other parts, the topological sensitivity field is set to a high value (a filtering strategy). Thus, when the topology is extracted, these parts are retained.

Note that this strategy does not depend on the material properties, or the loading scenario. However, in the current implementation of Pareto, it is assumed that all interfaces between parts are perfectly bonded.

■ **Example 11.1 CNC Frame Optimization**

1. From the *ParetoExamples* folder, select the *CNCMultiBody.stl* file; the model will be loaded as in Figure 11.2.

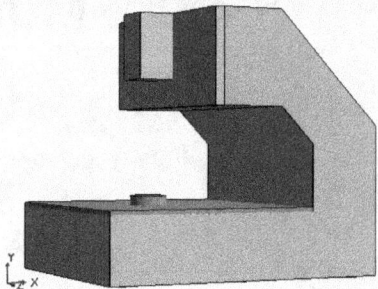

Figure 11.2: CNC multi-body model.

(R) The model consists of 3 bodies that are color coded; none of them have been assigned a material.

2. We will assign different materials to different bodies. First select the Part 0 (frame) as in Figure 11.3; select Cast Iron Grade 25 and apply.

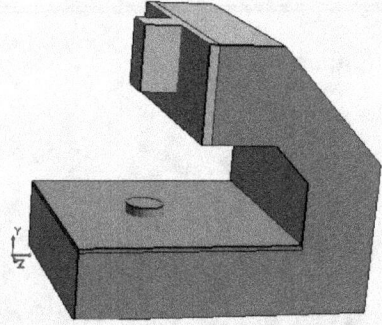

Figure 11.3: Selecting the frame and applying material properties.

3. Select Part 1 (base) as in Figure 11.4;
4. Select Alloy Steel, select *Do not optimize?* option and apply.

11.1 Multi-Body Optimization

ⓡ The last option is to ensure that material is not removed from the base plate during optimization.

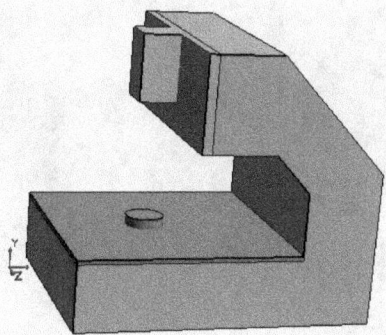

Figure 11.4: Selecting the base plate, and applying material properties.

5. Finally select Part 2 (tool holder) as in Figure 11.5.
6. Select Al 1060
7. Select *Do not optimize?* option and apply.

ⓡ Once again, the last option is to ensure that material is not removed from the base plate during optimization.

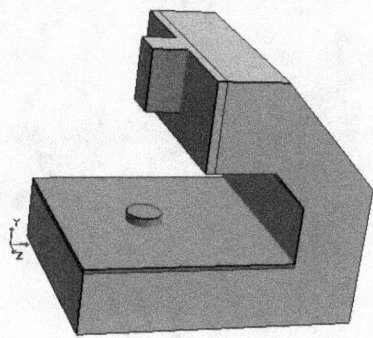

Figure 11.5: Selecting the tool holder, and applying material properties.

8. Once the material selection is complete, it will be reported in the graphics window as in Figure 11.6.

Model: CNCMultiBody
Length: 1.60 (meter)
Part 0: CastIronGr25
Part 1*: AlloySteel
Part 2*: Al1060
#LoadSets: 1

Figure 11.6: Multi-body material list (a * implies that the part must not be optimized).

9. Next, we will apply the restraints. Select the six cylindrical faces as in Figure 11.7, and apply fixed boundary condition.

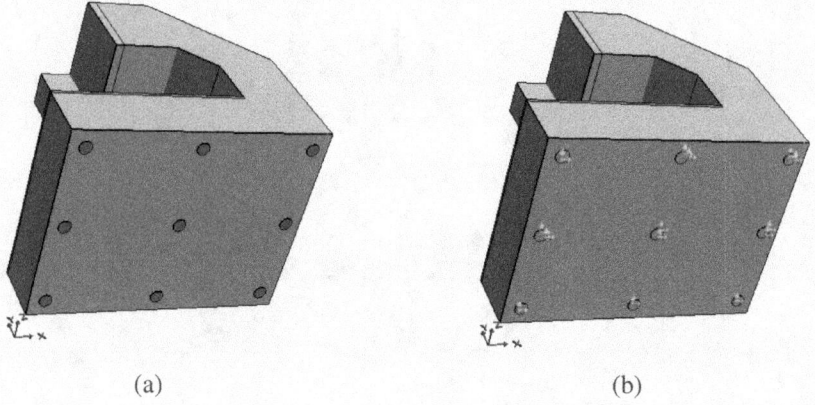

(a)   (b)

Figure 11.7: Applying the restraints.

10. We will apply two loads. Select the cylindrical face on the base plate as in Figure 11.8, and apply a normal force of 10,000 N.

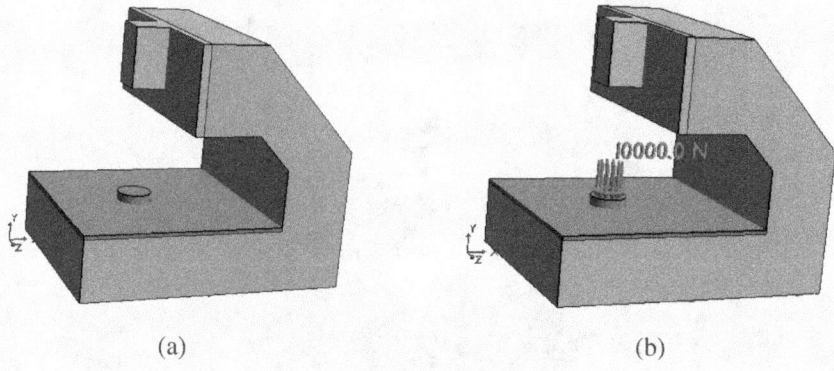

(a)   (b)

Figure 11.8: Applying the downward force of 10,000 N.

11. Select the bottom face on the tool holder as in Figure 11.9, and apply a normal force of 10,000 N.

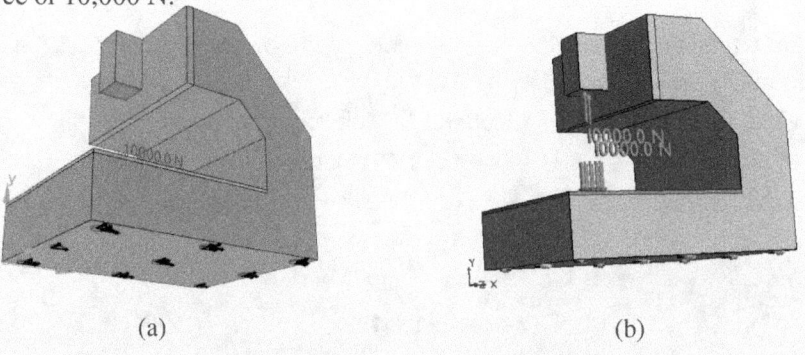

(a)   (b)

Figure 11.9: Applying the upward force of 10,000 N.

12. Under *FEA*, impose symmetry about z plane.
13. We will carry out a finite element analysis, using a fine quality mesh as in Figure 11.10. The maximum stress is around 0.6 MPa.

Figure 11.10: Finite element analysis with a fine quality mesh.

14. Under *TopOpt Constraints*:
    - The draw-direction is set to z-direction
    - Maximum displacement is 0.016 m.
15. The final step is to carry out topology optimization. The desired volume fraction is set to 0.35; the optimized topology is illustrated in Figure 11.11. Observe that the base-plate and the tool holder have not been optimized.

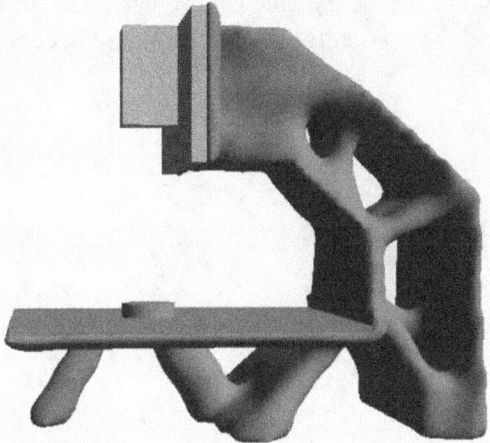

Figure 11.11: Optimized the CNC frame at 0.35 volume fraction.

16. The stress plot of the optimized topology is illustrated in Figure 11.12; the maximum stress is around 1.2 MPa.

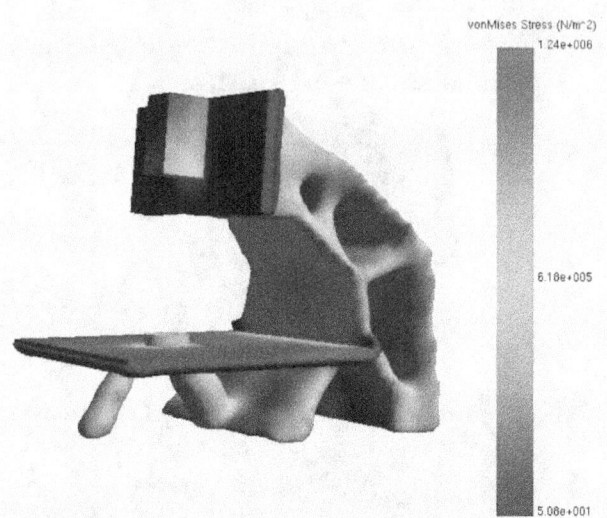

Figure 11.12: Stress result of the optimized topology.

17. Ssve the project as *CNCMultiBodyProject.prj*.

11.2 Multi-Material Optimization

In multi-material topology optimization (MMTO), one must simultaneously optimize the topology, and the distribution of various materials within the topology. While MMTO has been addressed by many researchers, the objective of this chapter is to pursue a MMTO method that is computationally efficient and can be generalized to variety of objectives. For multi-material topology optimization (MMTO), the compliance minimization problem can be generalized to:

$$\begin{aligned}
&\underset{\Omega_{k=1,...,M} \subset D}{minimize} \ f(\Omega, \mathbf{u}) = \mathbf{f}^T \mathbf{u} \\
&\Omega_i \cap \Omega_{i \neq j} = \emptyset \quad i,j = 1,..,M \\
&g_n(\Omega, \mathbf{u}) \leq 0 \quad n = 1,...,N \\
&\text{subject to} \\
&\mathbf{Ku} = \mathbf{f}
\end{aligned} \qquad (11.1)$$

where M is the number of materials and Ω_k is the topology for k th material to be computed.

Thus, the objective is to find the optimal distribution of M non-overlapping materials, within a given design space, that minimizes a specific objective and satisfies certain design constraints. The MMTO problem posed in Equation 11.1 assumes that every point within the design space has a distinct material associated with it (or is void). This differs from functionally graded material optimization, where a mixture of base materials is allowed.

11.2.1 Volume vs. Mass

In this section, we will compare two common formulations used in MMTO, namely volume-based and mass-based. The two objective give identical results in SMTO, since the density is constant over the entire domain. However these objectives are not necessarily equivalent in a non-homogeneous topology optimization, an important fact that is often neglected.

For simplicity, consider the following volume-constrained formulation to find the stiffest design at some volume fraction [11.2]:

$$\begin{aligned}
& \underset{\Omega_{k=1,\ldots,M} \subset D}{minimize} \; J(\Omega, \mathbf{u}) = \mathbf{f}^T \mathbf{u} \\
& |\Omega| \leq V^* \\
& |\Omega_k| \leq V_k^* \\
& \Omega_i \cap \Omega_{i \neq j} = \emptyset \quad i,j = 1,..,M \\
& g_n(\Omega, \mathbf{u}) \leq 0 \quad n = 1,...,N \\
& \text{subject to} \\
& \mathbf{Ku} = \mathbf{f}
\end{aligned} \qquad (11.2)$$

where J is compliance, $|\Omega|$ is the total volume, $|\Omega_k|$ is volume of kth material. Basically we are seeking the stiffest design while some volume constraints on each material phase as well as total volume are satisfied. Also, there are no overlaps between two different phases. Immediately we can see that in order to start the optimization process with the above formulation, we *must* have a very good approximation for the optimum volume of each material. However, this is clearly not a simple task for complex designs and loading conditions. Further, imposing these additional linear constraints will reduce the feasible design space to only a subset of the actual design space.

Now let us consider the following mass-constrained formulation:

$$\begin{aligned}
& \underset{\Omega_{k=1,\ldots,M} \subset D}{minimize} \; J(\Omega, \mathbf{u}) = \mathbf{f}^T \mathbf{u} \\
& \sum_{k=1,\ldots,M} \rho_k V_k \leq \bar{\mathcal{M}} \\
& \Omega_i \cap \Omega_{i \neq j} = \emptyset \quad i,j = 1,..,M \\
& g_n(\Omega, \mathbf{u}) \leq 0 \quad n = 1,...,N \\
& \text{subject to} \\
& \mathbf{Ku} = \mathbf{f}
\end{aligned} \qquad (11.3)$$

In other words, we are looking for the stiffest design while total mass of our design is less than some upper bound $\bar{\mathcal{M}}$. Observe that with Equation 11.3, we need not know the amount of each material prior to optimization. Further, in the absence of additional linear constraints, we are more likely to converge to the minimum compliance of all possible combinations of multiple material distributions. This is illustrated in Figure 11.13 borrowed from [11.3].

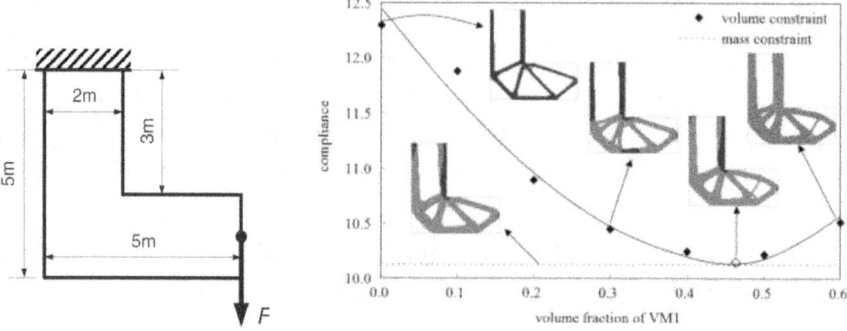

Figure 11.13: (a) Design Space and Boundary Condition (b) Comparison between results under mass and volume constraints with the equal mass [11.3].

Hence, we will present a formulation based on mass and later consider a more general objective function. Next, we will propose a multi-objective form of Equation 11.3 and discuss its advantages.

11.2.2 Multi-Objective Formulation

Most of the previous work on MMTO consider a single objective such as compliance and impose a constraint on mass. Alternatively, we can pose the problem in the following multi-objective form:

$$\begin{aligned}
& \underset{\Omega_{k=1,\ldots,M} \subset D}{\text{minimize}} \{J, \mathcal{M}\} \\
& \sum_{k=1,\ldots,M} \rho_k V_k = \mathcal{M} \\
& \Omega_i \cap \Omega_{i \neq j} = \emptyset \qquad i,j = 1,..,M \\
& \text{subject to} \\
& \mathbf{Ku} = \mathbf{f}
\end{aligned} \qquad (11.4)$$

Now we can solve this modified MMTO problem using the so-called Pareto frontier. This will be discussed in more details, but first, let us complete this section by generalizing Equation 11.4.

Generally, we consider any two conflicting quantities, namely cost (C) and performance. Typical quantities that represent cost include *weight* and *price (in dollars)* that must be minimized, while performance metrics include *stiffness*, *strength*, etc. that must be maximized. However, to be consistent with classic multi-objective optimization, where all quantities are minimized, we define inefficiency (I) as the inverse of performance. Consequently, the multi-material problem will be posed as:

$$\begin{aligned}
& \underset{\Omega_{k=1,\ldots,M} \subset D}{\text{minimize}} \{I, C\} \\
& \Omega_i \cap \Omega_{i \neq j} = \emptyset \qquad i,j = 1,..,M \\
& \text{subject to} \\
& \mathbf{Ku} = \mathbf{f}
\end{aligned} \qquad (11.5)$$

For example, for compliance (J) as Inefficiency and mass ($\mathcal{M}$) as Cost, we get the Equation 11.4.

11.2 Multi-Material Optimization

Assuming that the underlying design has been discretized using finite elements, and each element can be associated with a material of choice. Suppose we have a finite element with material k within the design. We will now consider a hypothetical swapping of the underlying material k to material m. Observe that the change in compliance (I is used here to denote inefficacy) is given by the first-order element sensitivity:

$$\Delta I_e^{k \to m} = \mathbf{u}_e^T \mathbf{K}_e^k \mathbf{u}_e - \mathbf{u}_e^T \mathbf{K}_e^m \mathbf{u}_e \tag{11.6}$$

As a special case, if the element is deleted, i.e., replaced with void, we have:

$$\Delta I_e^{k \to \emptyset} = \mathbf{u}_e^T \mathbf{K}_e^k \mathbf{u}_e \tag{11.7}$$

Similarly, as a special case of Equation 11.6, if a new element is inserted (in place of a void):

$$\Delta I_e^{\emptyset \to m} = -\mathbf{u}_e^T \mathbf{K}_e^m \mathbf{u}_e \tag{11.8}$$

Note that the above equation is consistent with the sensitivity expressions used in SIMP for compliance [11.4]. The element sensitivity in Equation 11.6 can obviously be generalized to other quantities of interest. Specifically, for any quantity of interest Q, the first-order sensitivity is given by:

$$\Delta I_e^{k \to m} = -\boldsymbol{\lambda}_e^T \mathbf{K}_e^k \mathbf{u}_e + \boldsymbol{\lambda}_e^T \mathbf{K}_e^m \mathbf{u}_e \tag{11.9}$$

Where $\boldsymbol{\lambda}$ is the adjoint field associated with the quantity of interest. Thus, there is no fundamental restriction of the proposed method to compliance problems. The specific expression for the adjoint depends on the quantity of interest; for example, for the p-norm stress, an expression for the adjoint is derived in [11.1].
Correspondingly, the change in cost can also be computed; for example, if the cost is the mass function:

$$\Delta C_e^{k \to m} = \rho^m V_e - \rho^k V_e \tag{11.10}$$

where ρ denotes (real) material density and V_e denotes volume of an element.

11.2.3 Algorithm

Our objective is to trace the Pareto curve involving these two quantities with cost (C) on the x axis and inefficiency (I) on the y axis. For simplicity, let us for now assume that the initial optimum design is known; for instance the most expensive material also performs best and there are no materials with the same cost. We shall discuss finding the initial Pareto design in case of ambiguities. The proposed algorithm will start with a topology with the highest cost and optimized material distribution and trace Pareto front through a series of fixed-point iterations.

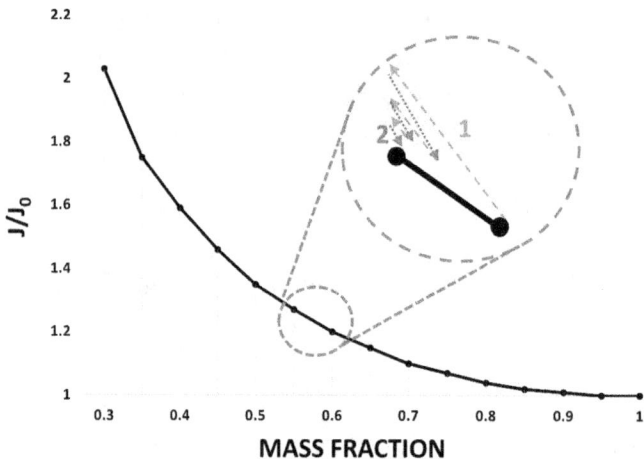

Figure 11.14: Tracing Pareto curve, and the two sub-steps.

To ensure that each optimization step results in a Pareto-optimal design, each step consists of two sub-steps, where (1) the cost function will be reduced by a small decrement (sub-algorithm 1), followed by (2) a reduction in inefficiency (sub-algorithm 2); these two steps are illustrated in Figure 55. By repeating these two steps, we show that the Pareto curve can be generated, and associated topologies computed, in an efficient and unambiguous manner.

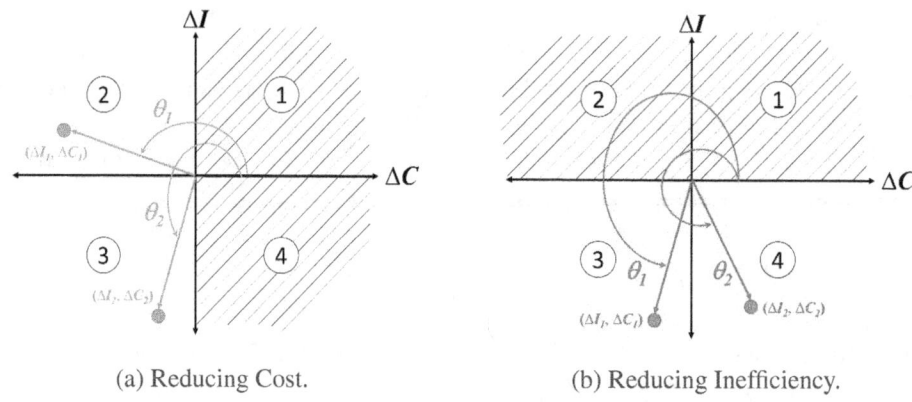

(a) Reducing Cost.   (b) Reducing Inefficiency.

Figure 11.15: Ranking parameters at the two sub-steps.

1) Reducing Cost

Now consider the cost-inefficacy plot of Figure 11.15a. In this sub-step, our objective is to reduce cost, i.e., we swap materials only if $\Delta C_e^{k \to m} < 0$. The change in inefficacy can be either positive or negative Therefore, only quadrants 2 and 3 in Figure 11.15a are acceptable; quadrant 3 is preferable, since both cost and inefficacy are reduced. Thus the angle θ must

11.2 Multi-Material Optimization

be maximized, i.e.,

$$\underset{m}{\text{maximize}}\; \theta_e^{k \to m} \equiv \frac{\Delta I_e^{k \to m}}{\Delta C_e^{k \to m}}$$
$$\text{subject to}$$
$$\Delta C_e^{k \to m} < 0 \tag{11.11}$$

or equivalently:

$$\underset{m}{\text{minimize}}\; -\theta_e^{k \to m} \equiv \frac{-\Delta I_e^{k \to m}}{\Delta C_e^{k \to m}}$$
$$\text{subject to}$$
$$\Delta C_e^{k \to m} < 0 \tag{11.12}$$

Hence, in the first sub-step, for each element, we find the material m that gives the lowest value of $-\theta_e^{k \to m}$.

2) Reducing Inefficiency

Now consider the cost-inefficacy plot of Figure 11.15b. In this sub-step, our objective is to reduce cost, i.e., we swap materials only if $\Delta I_e^{k \to m} < 0$. The change in cost can be either positive or negative Therefore, only quadrants 3 and 4 in Figure 11.15b are acceptable; considering proper normalization, we try to stay as close as possible to $\theta_e^{k \to m} = \frac{3\pi}{2}$.

$$\underset{m}{\text{minimize}}\; |\theta_e^{k \to m} - \frac{3\pi}{2}| \equiv \text{sign}(\Delta C_e^{k \to m}) \frac{\Delta I_e^{k \to m}}{\Delta C_e^{k \to m}}$$
$$\text{subject to}$$
$$\Delta I_e^{k \to m} < 0 \tag{11.13}$$

Considering the above discussion, the details of algorithms are as follows:

> **Algorithm 11.2.1 Main algorithm**
> The main algorithm is illustrated in Figure 11.16 and each of the steps is described below.
> 1. Start with an initial design with $\Omega = D$ and maximum cost (example, maximum weight).
> 2. The cost C is reduced by ΔC, either by removing material or replacing the most costlier material one with the less expensive one (see sub-step 1 for details).
> 3. Inefficiency I is reduced while maximum of N number of elements can be modified. The cost can increase or decrease. (see sub-step 2 for details)
> 4. Check if the converged design is acceptable:
> 5. If Step-3 fails, reduce ΔC and repeat
> 6. If the design has converged to the desired cost C, optimization is terminated
> 7. (Else) ΔC is reinitialized, and algorithm returns to step 1.

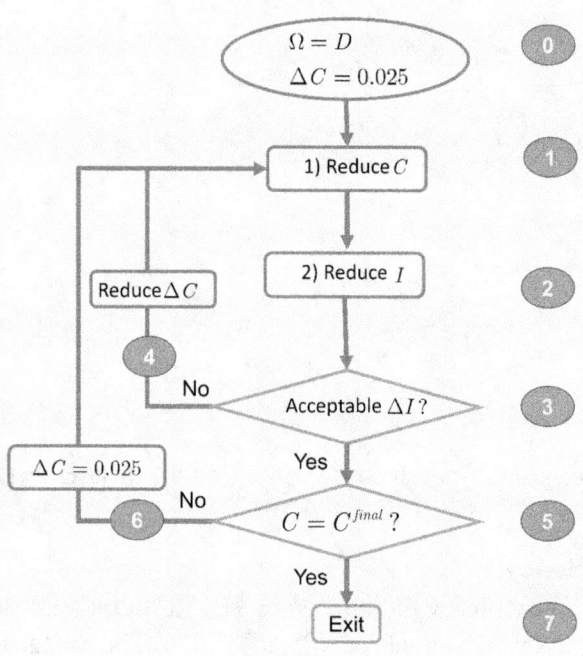

Figure 11.16: Main algorithm.

Algorithm 11.2.2 **Sub-step 1: Reduce Cost C**
The sub-algorithm 1 is illustrated in Figure 11.17, and each of the steps is described below.
1. We first perform a finite element analysis and compute the inefficiency for each element.
2. Next, for *each element* we find the ranking parameter $r(e)$.
3. In the next step, we sort the array r in an increasing order while keeping track of corresponding material and element.
4. Initialize counter i and the current reduction in C in this step (δC) to zero.
5. Replace elements with a new material or void, accordingly.
6. Update values of δC and i.
7. Check if we have reached the allowed ΔC.
8. (Yes) Update C and go to sub-step 2.
9. (No) return to 1.5

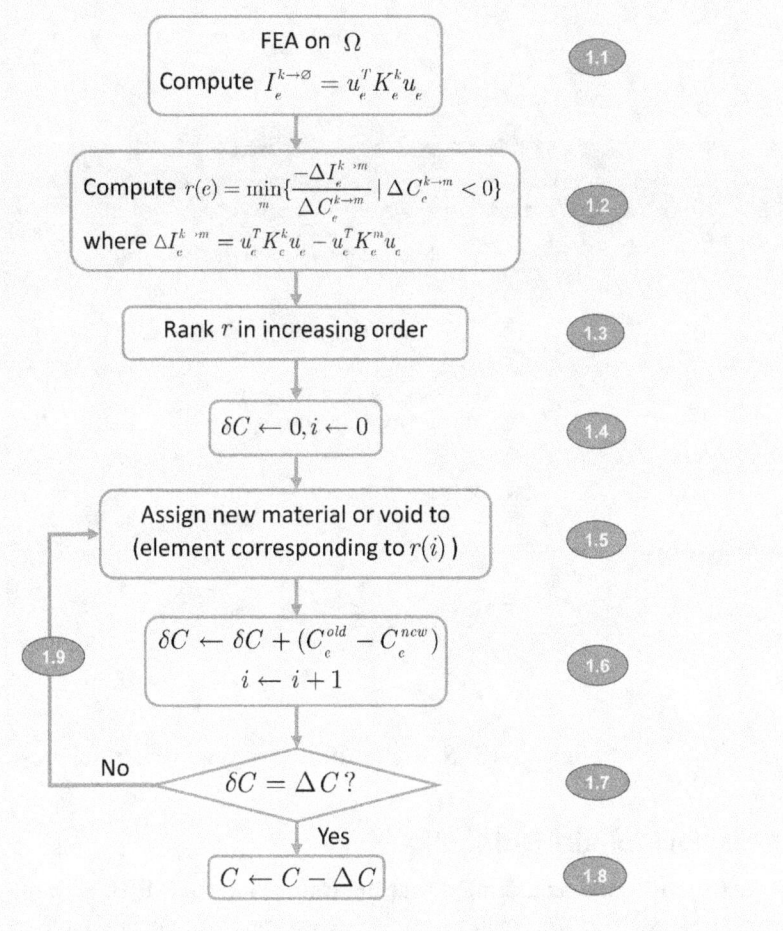

Figure 11.17: Sub-algorithm 1, reducing C.

Algorithm 11.2.3 Sub-step 2: Reduce Inefficiency at Constant C
Figure 11.18 shows the sub-algorithm 2, and each of its steps is explained herein.
1. We first perform a finite element analysis and compute the inefficiency for each element
2. Next, for each element we find the ranking parameter $r(e)$.
3. In the next step, we sort the array r in an increasing order while keeping track of corresponding material and element.
4. Initialize counter i and the current reduction in C in this step (δC) to zero.
5. Replace elements with a new material or bring back void elements, accordingly.
6. Update values of (δC) and i.
7. Check if we have reached the allowed number of element changes N.
8. (Yes) Update C and go to sub-step 2
9. (No) return to 2.5

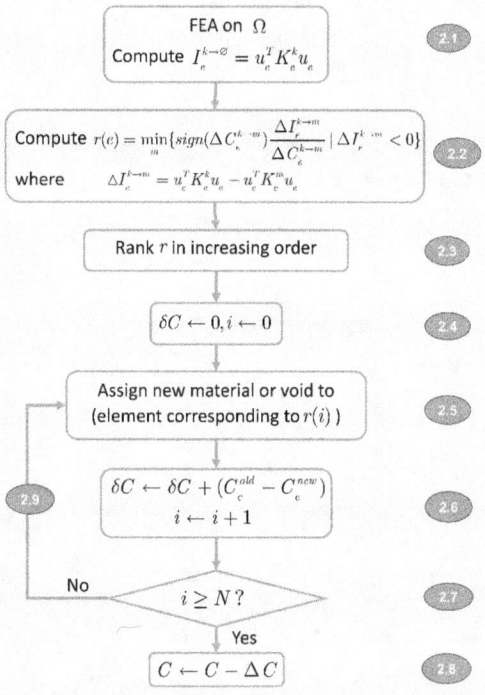

Figure 11.18: Sub-algorithm 2, reducing inefficiency.

11.2.4 Initial Material Distribution

In the last section, it was assumed that the initial material distribution is uniquely known as *a priori*. However, this may not be the case when materials have the similar costs or directional dependence, e.g. composites. Finding initial optimal design is addressed in this section, where the material distribution (and possibly geometry) is optimized while the cost is remained a constant.

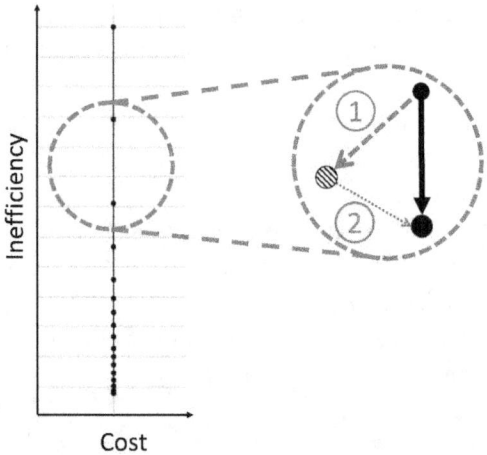

Figure 11.19: Optimizing design at the same cost.

11.2 Multi-Material Optimization

The algorithm is essentially a special case of the main algorithm of Figure 11.16, where we begin with some sub-optimal distribution of material and successively apply sub-algorithms 1 and 2 with similar ΔC, until we find a local optimum which lies on the Pareto frontier.

11.2.5 Numerical Validations

In this section, we demonstrate the validity of the proposed method through several 3D examples. In all numerical experiments, the decrement in the cost function for Pareto tracing is initialized to 0.025. All dimension are in meters, unless otherwise noted. In all tables, E, v, and ρ denote Young's modulus, Poisson ratio, and density, respectively.

L-Bracket: Single and Two-Material Design

First, we consider the L-bracket illustrated in Figure 11.20a and compare the optimized results for: (1) a single material A, and (2) two materials A and B, whose properties are summarized in Table 11.1. The mass serves as the cost function while compliance is the measure of inefficiency; 20,000 elements are used for both experiments.

Table 11.1: Material Properties of A and B.

Material	$E(GPa)$	v	$\rho(Kg/m^3)$
A	170	0.3	7100
B	70	0.33	2700

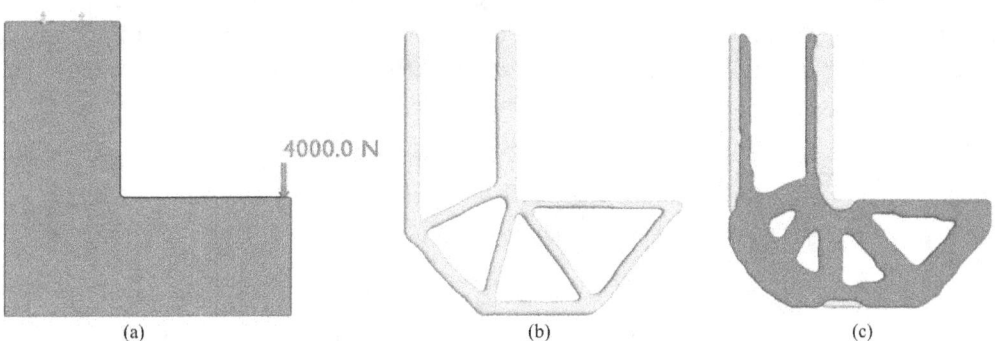

Figure 11.20: (a)L-bracket geometry and loading condition, (b) design at 30% cost using only material A and (c) design at 30% cost using materials A and B.

The Pareto curve is generated for up to 70% reduction in mass. As stated earlier in the algorithm, we start with the heaviest design (all A) and optimize the topology and material distribution. After the optimization process is complete, we obtain the two Pareto curves illustrated in Figure 11.21. As expected, for a given weight fraction, the two-material design is less compliant (stiffer) than the corresponding single material design.

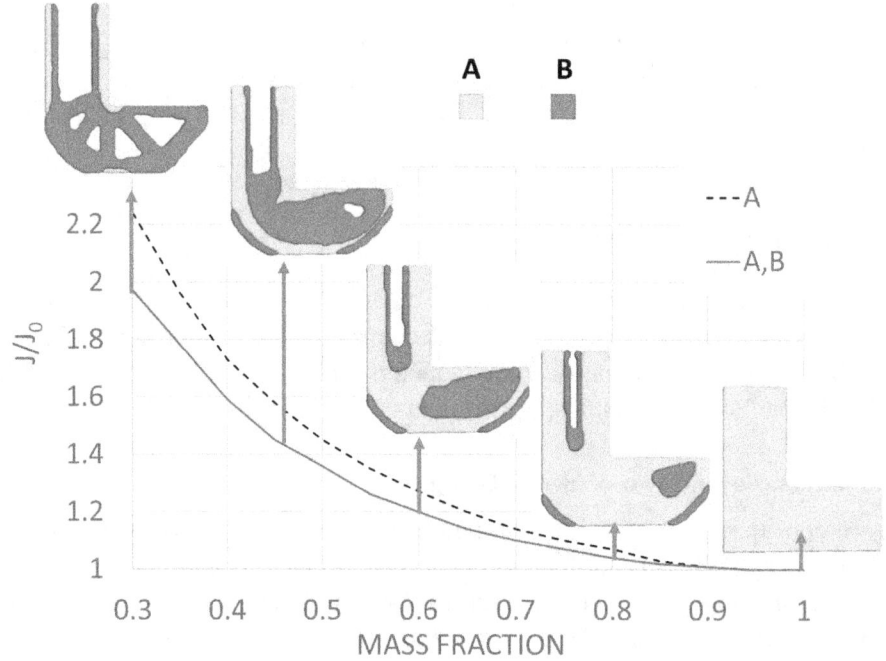

Figure 11.21: The Pareto curves for single material (A) and two materials (A and B).

Figure 11.22 illustrates the number of deflated CG iterations for each of the 32 FEAs during optimization. Figure 64a corresponds to single-material, while Figure 64b corresponds to multi-material. As one can observe, the iterations increase moderately in both scenarios; this can be attributed to the presence of thin/slender structures in the topology. Both experiments required 32 finite element operations to complete. Thus, for this example, the cost of multi-material design is almost exactly the same as the single-material design.

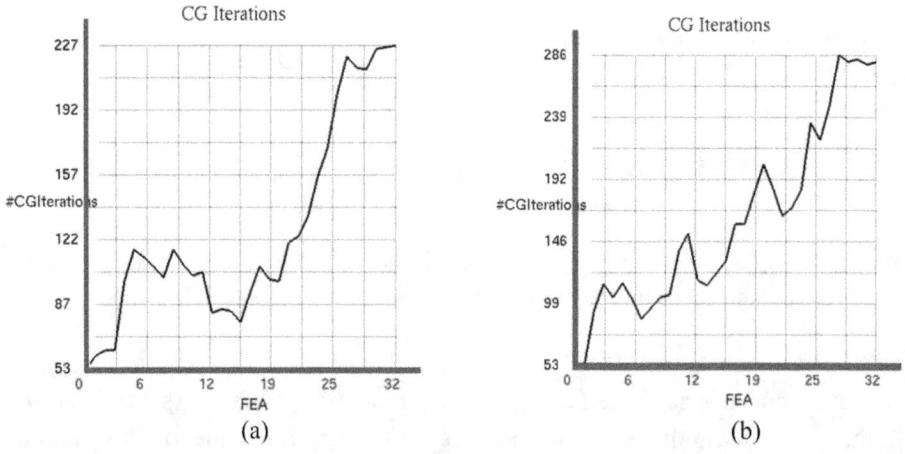

Figure 11.22: CG iterations for the L-Bracket with (a) single material A, (b) two materials A and B.

11.2 Multi-Material Optimization

MBB Structure: Mesh Independency

In this example, we study the impact of mesh size using the classic MBB structure illustrated in Figure 11.23. The structure is loaded with 30 units at the center, and 15 units on either side of the center; the material properties are given in Table 11.1.

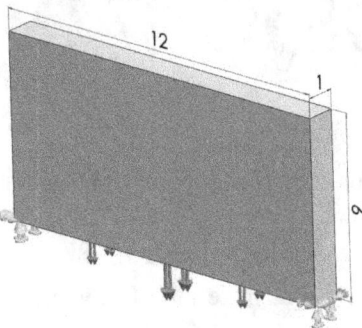

Figure 11.23: MBB structure.

Figure 11.24 illustrates the Pareto curves with 10,000 and 40,000 elements; as one can observe, the two curves are almost identical, suggesting that the method is insensitive to mesh discretization.

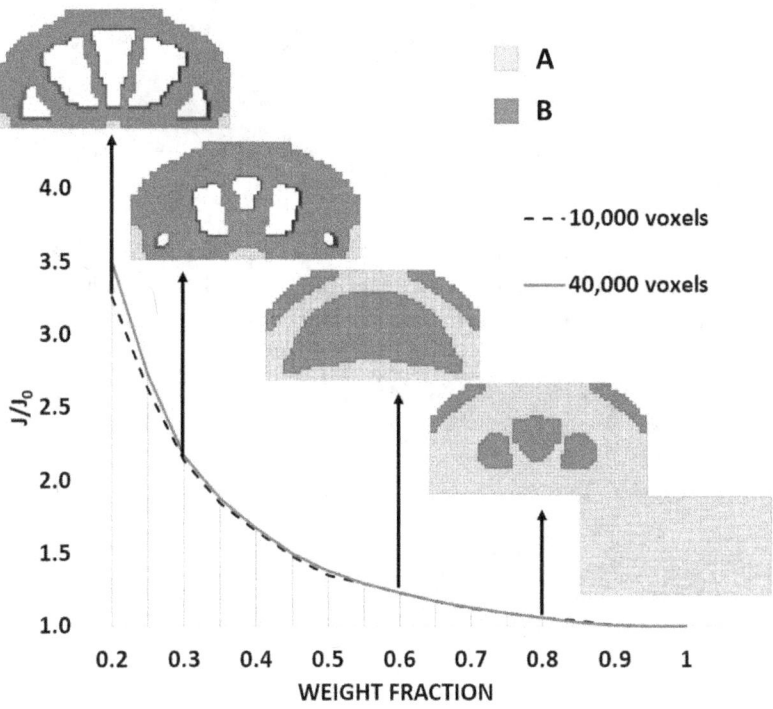

Figure 11.24: Pareto curves for 10,000 and 40,000 elements.

The final topologies are illustrated in Figure 11.25. Note that as the mesh size increases, the design details are a bit refined as expected. Yet the distribution of the material within the two designs are quite similar.

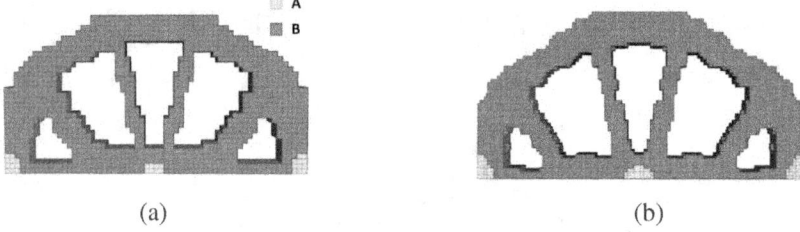

(a)   (b)

Figure 11.25: MBB structure with 10,000 and 40,000 elements at weight fraction of 0.2.

Cantilevered Beam: Three-Material Pareto Curve

In this experiment, we consider three materials. The geometry and boundary conditions are illustrated in Figure 11.26. The design is discretized using 30,000 elements, i.e. 101,400 degrees of freedom.

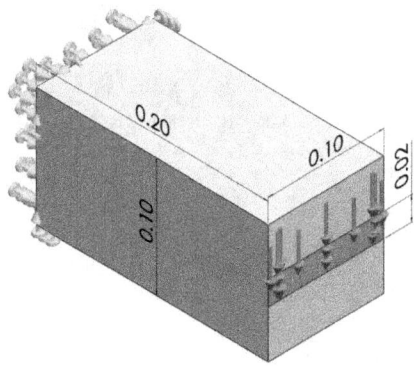

Figure 11.26: Cantilevered beam.

The material properties are summarized in Table 11.2. We solve the MMTO problem for 3 different scenarios: (1) pure A (single material), (2) A and B (two materials) and (3) A, B, and C (three materials).

Table 11.2: Material Properties of A, B and C.

Material	$E(GPa)$	v	$\rho(Kg/m^3)$
A	380	0.2	19250
B	210	0.3	7800
B	110	0.25	4390

Beginning with all A initial design, the objective is to reduce mass by 70% while keeping the design as stiff as possible. Figure 11.27a shows the optimized design using only material A, Figure 11.27b is optimized design with materials A and B. Figure 11.27c illustrates optimized designs using A and C. Finally, Figure 11.27d illustrates the optimized design at 0.3 mass fraction using all three materials.

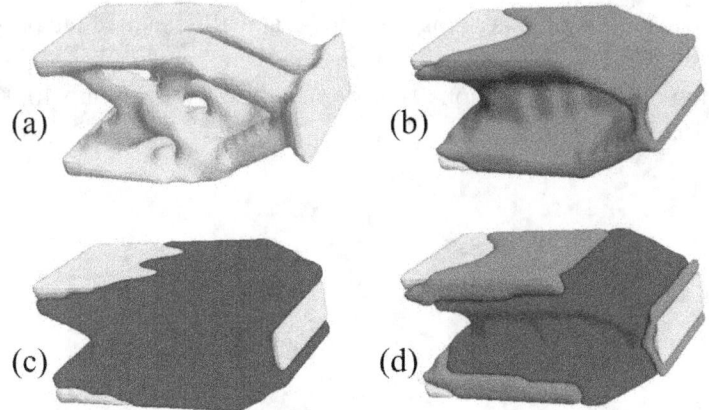

Figure 11.27: Optimized cantilever beam at 0.3 mass fraction, (a) only A ,(b) A and B, (c) A,C, and (d) A, B, and C.

Figure 11.28 illustrates the 4 Pareto curves, and the topologies for the third scenario. As one can observe, moving from one material to two-materials results in a significant improvement of 27% for both cases of A-B and A-C, and adding the third-material further improves the design by 29%.

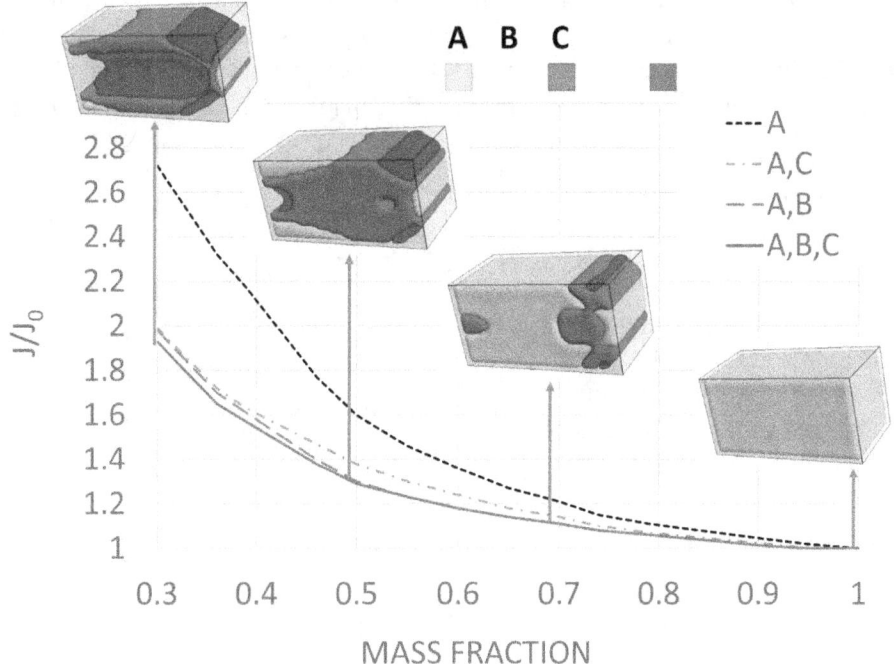

Figure 11.28: Effect of number of materials on Pareto Curve.

It is also interesting that up to 50% mass reduction, A-B results are identical to those of

A-B-C and only in lower mass fractions does the third material improve the result.

Mount-Bracket: a case study

This example focuses on demonstrating robustness and efficiency of the proposed method via a more complex and large scale design. Consider the mount bracket of Figure 11.29 and its corresponding boundary conditions. The domain is discretized into 40,000 hexahedral elements with 163,929 dof.

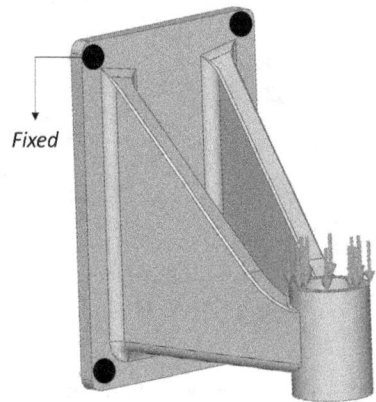

Figure 11.29: Mount bracket, geometry and boundary conditions.

There are many metallic filaments available for FDM. For this example, let us consider Copper-PLA and Aluminum-PLA filaments. Neglecting anisotropic behavior due to lack of data, the effective isotropic material properties for these materials can be found via applying the rule of mixture and are according to Table 11.3. Obviously, copper filament is both stiffer and heavier.

Table 11.3: Material Properties of Copper-PLA and Aluminum-PLA.

Material	$E(GPa)$	v	$\rho(Kg/m^3)$
Copper-PLA (40%-60%)	45.2	0.33	4340
Aluminum-PLA (40%-60%)	28.8	0.33	1836

Given these material choices and beginning with the stiffest and heaviest initial design, the objective is to reduce mass by 60%. Figure 11.31 illustrates cross-sectional views of the design of Figure 11.30b. Using both materials reduces compliance of final design by about 13%.

Figure 11.32 illustrates the Pareto fronts for single-material TO with Copper-PLA and bi-material TO with both materials in Table 11.3. Further, as stated in section 4.4.6, tracing the Pareto curve need not begin at the heaviest design. For the two materials of Table 11.3 we can generate a random material distribution, which results in a sub-optimal design at 0.7 mass

11.2 Multi-Material Optimization

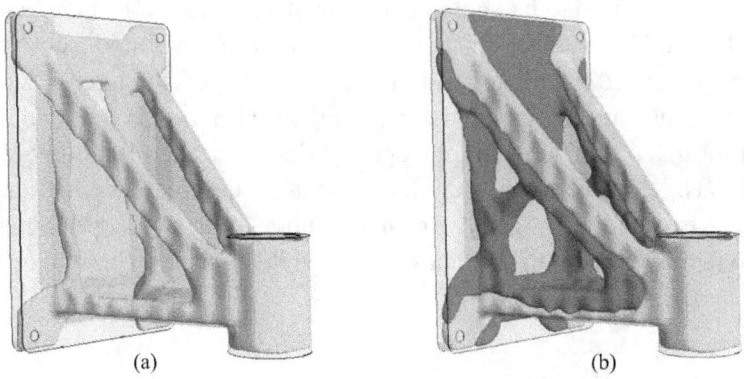

Figure 11.30: Mount bracket, optimized design at mass fraction of 0.4 using (a) only A and (b) A and B.

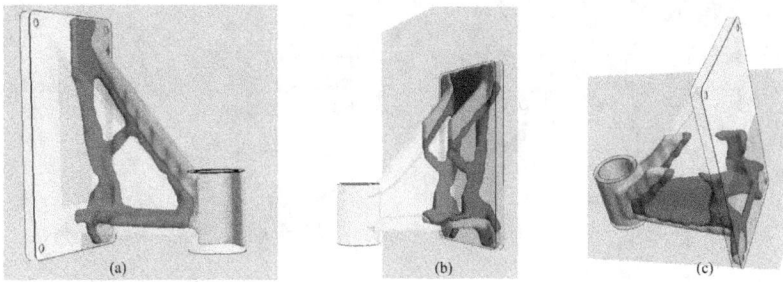

Figure 11.31: Mount bracket cross-sectional views.

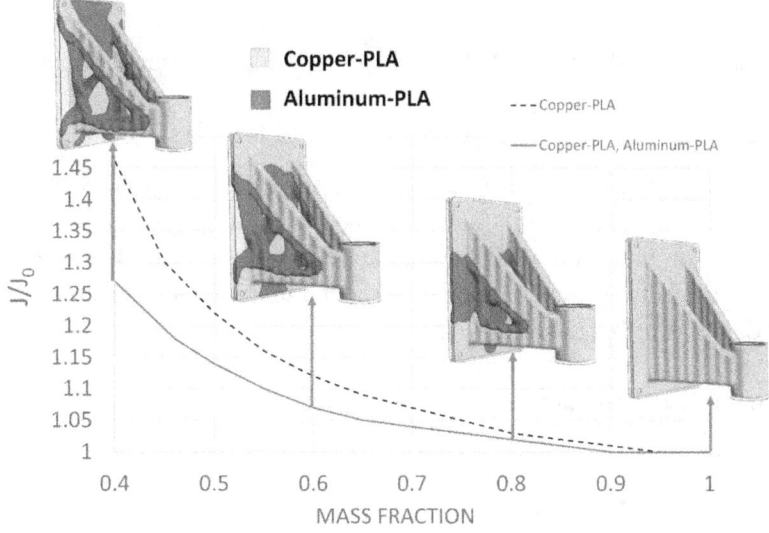

Figure 11.32: Mount bracket Pareto frontiers for copper-PLA and Aluminum-PLA filaments.

fraction. Thus, we first need to find optimal material distribution at 0.7 mass fraction through algorithm of Figure 11.19. Once this Pareto-optimal design is found, subsequent optimal designs can be generated via the algorithm of Figure 11.16. The process is demonstrated and compared against original Pareto tracing in Figure 11.33.

Observe that for this particular example, optimal design at mass fraction of 0.7 is similar for both methods. However, it is well-known that for complex geometries and loading conditions, there exist multiple (locally) optimal solutions with similar performance. It might be possible to rapidly explore the design space through efficiently combining algorithms of Figure 11.16 and Figure 11.19.

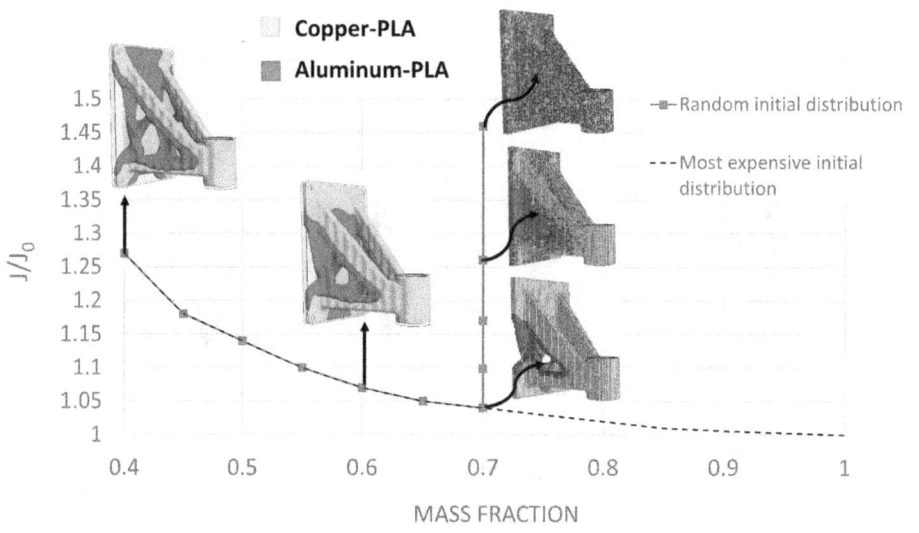

Figure 11.33: Mount bracket with random initial material distribution.

Run times and Memory Requirements

For all the experiments presented in this section, Table 11.4 summarizes the degrees of freedom, the target weight (or cost), total run-time, and required memory. All experiments were conducted on an Intel Core i7 CPU running at 3.4 GHz with 8GB of memory. Observe that all of the optimizations are completed in about a minute, and use the limited memory (in the order of a tens of Mega-Bytes).

Table 11.4: Summary of computational costs.

Example	#DOF	Target weight	Time	Memory (MB)
L-bracket	41,700	30%	2 min.	40
MBB	36,300	20%	2 min	30
Beam	34,000	20%	2 min	30
Mount Bracket	163,929	40%	3 min	60

(R) The Pareto software release currently does not expose the multi-material optimization framework; this will be included in a future release of Pareto.

11.3 Exercises

Exercise 11.1 Consider the knuckle assembly in Figure 11.34; the geometry is available as KnuckleAssembly.stl in ParetoExamples folder. Treating this as a multi-body problem, optimize the knuckle (but not the shaft) for the lowest possible volume fraction. Assume that both components are made of alloy steel, mesh to be of medium quality, fixed faces are retained, and x and z symmetries apply. No other topological constraints apply. How does the design differ from the one computed earlier where the shaft was not included? Save the project.

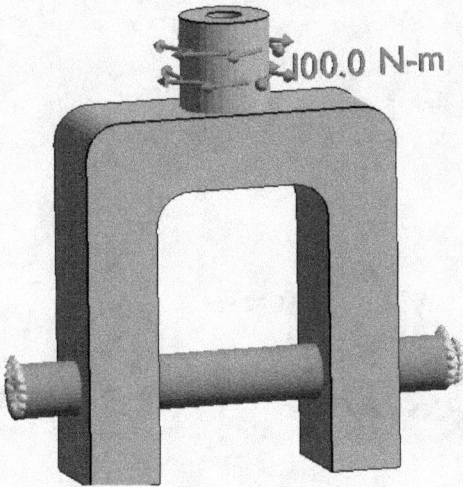

Figure 11.34: A multi-body optimization problem where the knuckle must be optimized.

Exercise 11.2 In the above the knuckle problem, assume that the shaft is made of aluminum 1060 (while the knuckle is alloy steel). Compare against the previous design.

Exercise 11.3 Repeat the above two exercises where the torque is replaced by a tensile loading of 13,000 N.

Exercise 11.4 Now consider both the torque and tensile loading for the knuckle problem, i.e., carry out a multi-load, multi-body optimization.

Exercise 11.5 Consider the spindle mount problem in Figure 12.11. Treating this as a multi-body problem, optimize just the spindle mount.

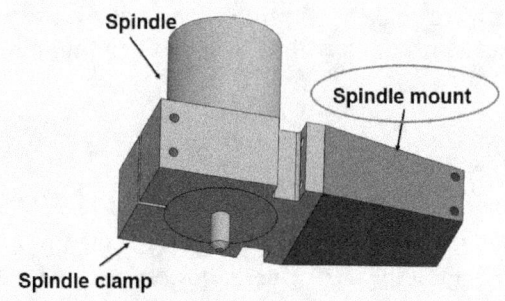

Figure 11.35: A multi-body optimization problem where only the spindle-mount must be optimized.

Exercise 11.6 Consider the two-load problem illustrated in Figure 11.36. All other problem parameters being the same. Optimize the spindle mount using a multi-load, multi-body framework.

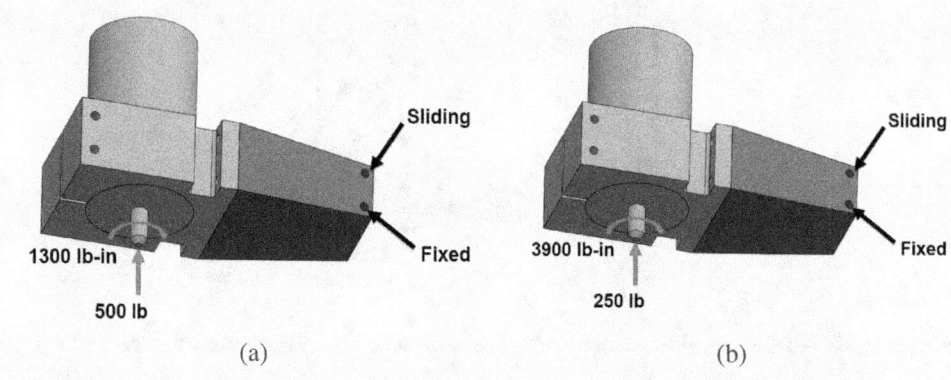

Figure 11.36: Multi-load problem for the spindle mount.

11.4 References

Reference 11.1 K. Suresh and M. Takalloozadeh, *Stress-constrained topology optimization: a topological level-set approach*, Struct Multidisc Optim, vol. 48, no. 2, pp. 295–309, Mar. 2013.

Reference 11.2 M. Y. Wang and X. Wang, *'Color' level sets: a multi-phase method for structural topology optimization with multiple materials*, Computer Methods in Applied Mechanics and Engineering, vol. 193, no. 6–8, pp. 469–496, Feb. 2004.

Reference 11.3 T. Gao and W. Zhang, *A mass constraint formulation for structural topology optimization with multiphase materials*, International Journal for Numerical Methods in Engineering, vol. 88, no. 8, pp. 774–796, Nov. 2011.

11.4 References

Reference 11.4 G. I. N. Rozvany, *A critical review of established methods of structural topology optimization*, Struct Multidisc Optim, vol. 37, no. 3, pp. 217–237, Jan. 2009.

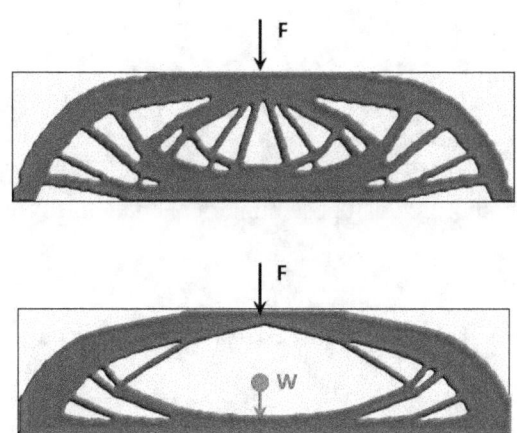

12. Gravity Loading

In this chapter, we will learn how to include gravity loads, and its impact on topology optimization.

12.1 Body Forces

Gravity loading is a special case of a body force, that must often be accounted for during topology optimization. Towards this end, consider the structural analysis problem:

$$\mathbf{Ku} = \mathbf{f} \tag{12.1}$$

where $\mathbf{K}$ is the stiffness matrix, $\mathbf{f}$ is the total force and $\mathbf{u}$ is the unknown displacement field. Previously the force $\mathbf{f}$ consisted of just the surface (traction) force, whose sensitivity with respect to topological changes vanishes, leading to the sensitivity equation:

$$\mathbf{K}'\mathbf{u} + \mathbf{K}\mathbf{u}' = 0 \tag{12.2}$$

In the presence of a body force, we have:

$$\mathbf{f} = \mathbf{f}_S + \mathbf{f}_N \tag{12.3}$$

The sensitivity of the body force does not vanish, leading to

$$\mathbf{K}'\mathbf{u} + \mathbf{K}\mathbf{u}' = \mathbf{f}'_N \tag{12.4}$$

The expression for the body force sensitivity depends is simply given by

$$\mathbf{f}'_N = \mathbf{b} \tag{12.5}$$

where $\mathbf{b}$ is the force per unit volume. This is appended to the classic topological sensitivity expression discussed in Chapter 4, and the same optimization algorithm can be applied to find optimal topologies.

An advantage of the Pareto method in the presence of body force is that, since the topology is always connected, the presence of a body force will not lead to a singularity. On the other hand, for methods where the topology can get disconnected during optimization, the presence of a body force can pose numerical challenges.

12.2 Examples

We will now consider a few examples to illustrate topology optimization under a body force.

▪ Example 12.1 MBB Problem

1. Load the pre-defined project: *MBBProject.prj* from the *ParetoExamples* folder, as in Figure 12.1.

   ® The initial design is a thin plate-like structure that is fixed at one end, and has a sliding constraint on the other. A nominal force is applied at the center.

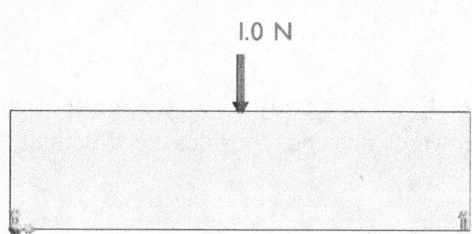

Figure 12.1: MBB Project: loading and restraints.

2. Using a fine quality mesh, carry out a finite element analysis as in Figure 12.2; the maximum stress is 32 ksi.

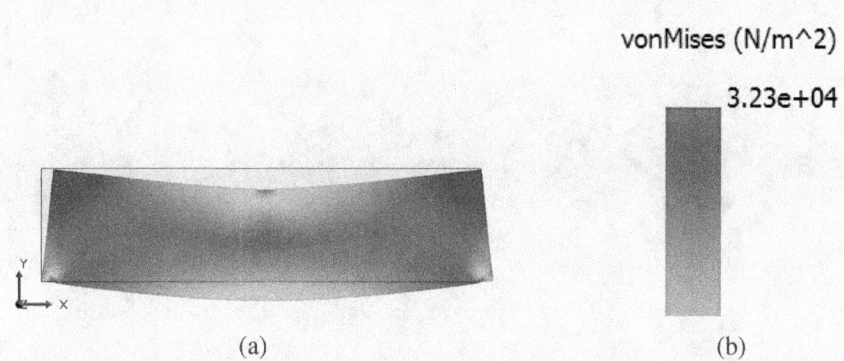

Figure 12.2: Finite element analysis.

3. Next, carry out optimization with the parameters specified in the project file.
4. The optimized design at 0.6 volume fraction is illustrated in Figure 12.3.

Figure 12.3: Optimized design without gravity loading.

5. Next, we will apply an additional gravity loading. Under the *Body Force* menu, set the y component to -9.81 (gravity), and apply.
6. The resulting display is illustrated in Figure 12.4.

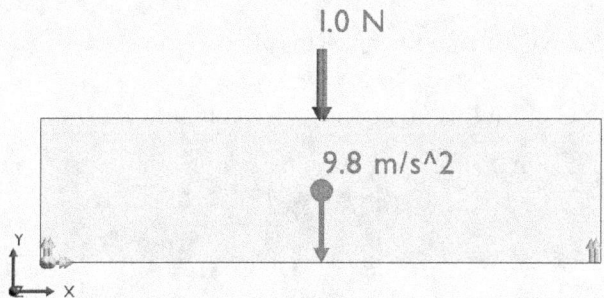

Figure 12.4: MBB problem with gravity.

7. Using a fine quality mesh, carry out a finite element analysis as in Figure 12.5; the maximum stress is 435 ksi (compare against Figure 12.2).

Figure 12.5: Finite element analysis with gravity loading.

8. Next, carry out optimization with the same parameters as specified in the project file. The optimized design at 0.6 volume fraction is illustrated in Figure 12.6.

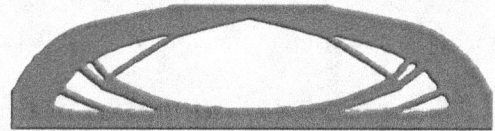

Figure 12.6: Optimized design with gravity loading.

(R) Observe the differences with respect to the topology in Figure 12.3. With the presence of gravity, the material in the center is penalized, and therefore eliminated during optimization.

We will next reconsider the CNC frame design, with gravity loading.

■ Example 12.2 CNC Frame Design Revisited

1. Upload the CNCMultiBodyProject.prj; see Figure 12.7.

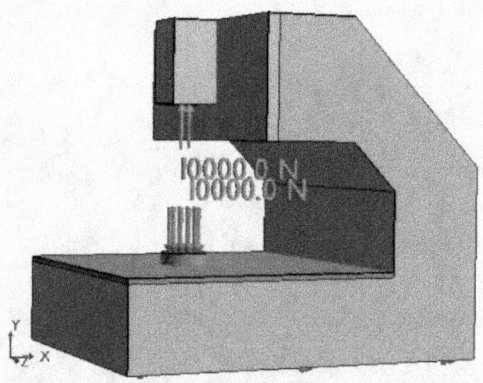

Figure 12.7: CNC multi-body problem.

12.2 Examples

2. The optimization parameter settings are as imposed previously.
3. Next, we will add gravity loading as before; see Figure 12.8.

   ⓡ Note that gravity applies to all three bodies. Further, the force that each member experiences depends on the density of the material.

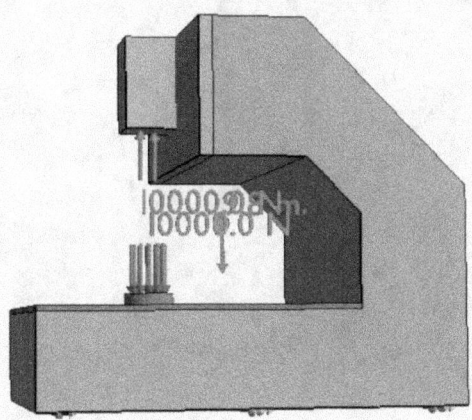

Figure 12.8: Restraints, forces, and gravity loading.

   ⓡ Observe that the center of mass might be outside the solid body.

4. Carry out the optimization with the same set of constraints as before.
5. The optimized frame design, with gravity loading, is illustrated in Figure 12.9b. One can observe significant differences compared to the topology in Figure 12.9a.

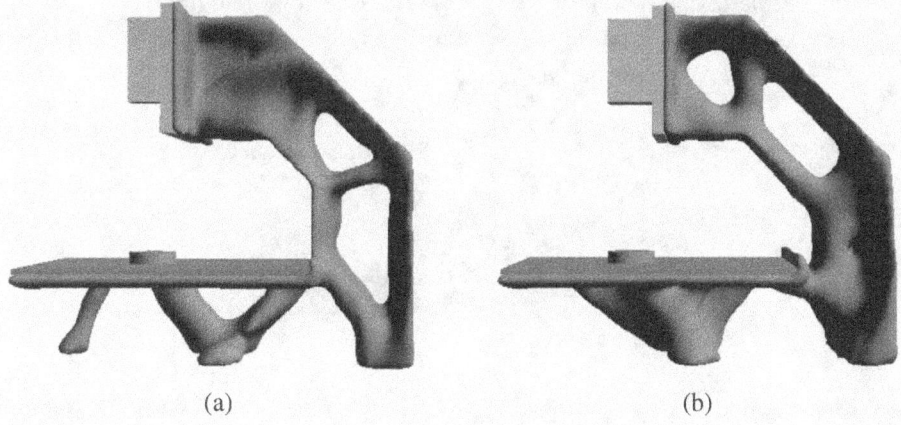

(a)   (b)

Figure 12.9: Optimized frame at 0.35 volume fraction, (a) without gravity, and (b) with gravity.

12.3 Exercises

Exercise 12.1 Consider the table problem in Figure 12.10. Assuume that the material is ABSPlastic and all four corners are fixed. Using a fine quality mesh, ptimize the design, without gravity, for a desired volume fraction oif 0.01 . Now apply a gravitional load of -388.22 (in imperial units) along y direction, repeat the exercise, and compare the results.

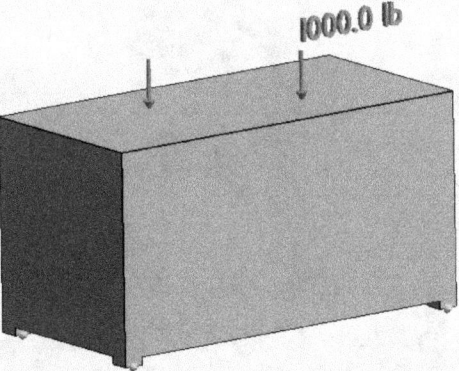

Figure 12.10: A table design problem.

Exercise 12.2 Repeat the above exercise with the external force set to 100 lbs, and 5000 lbs.

Exercise 12.3 Consider the spindle mount problem in Figure 12.11. Treat this as a multi-body problem, and include gravity. Optimize the spindle mount.

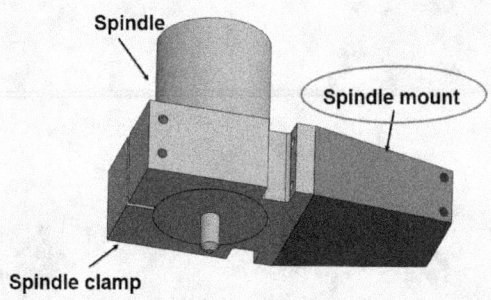

Figure 12.11: A multi-body optimization problem where only the spindle-mount must be optimized.

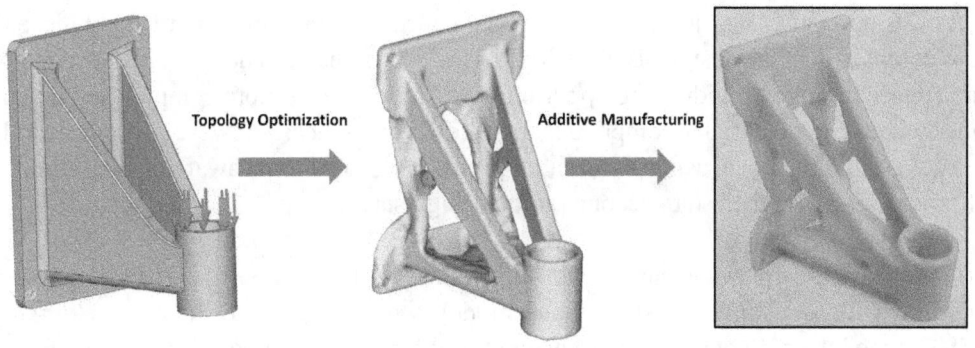

13. Additive Manufacturing

Additive manufacturing (AM) refers to a class of manufacturing processes through which parts are fabricated by material addition. The growing interest in AM stems from its ability to fabricate highly complex parts, a feat that was not possible through traditional manufacturing techniques such as laser-cutting, casting, or milling. Further, AM is well-suited for a great variety of applications across multiple disciplines from engineering to biomedical to art and architecture that require (and/or):

- Mass-customization
- Small-batch production
- On-spot fabrication
- Creative design

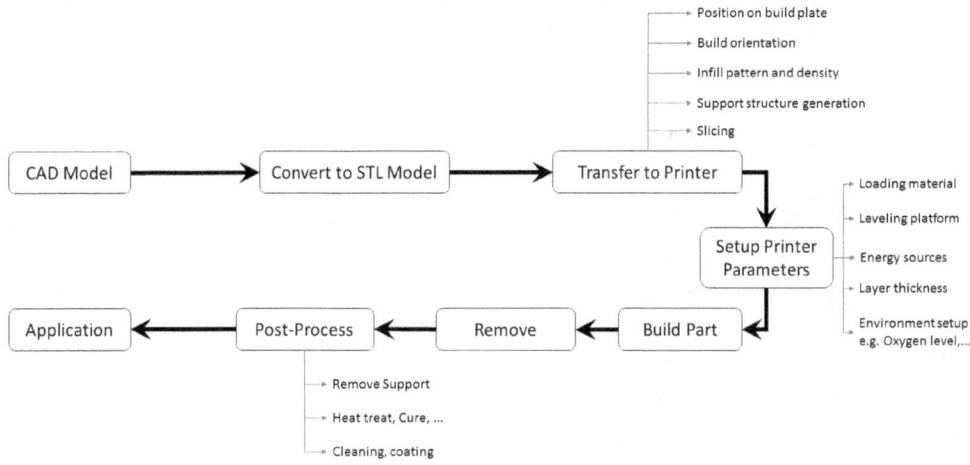

Figure 13.1: Workflow of design for Additive manufacturing.

As was previously discussed, additive manufacturing technologies are relatively insensitive to complexity of the design. Thus, it is widely accepted that AM has greatly expanded the design space through providing the opportunity to fabricate ever more complex and organic shapes which can potentially comprise many free-form surfaces. However, to successfully fabricate a functional part using a particular process, we need to be aware of limitations of that process and take them into account at the design stage.

Figure 13.1 illustrates the current workflow of design for additive manufacturing, where beginning with any representation of the model, we first generate the *STereoLithography*(STL) model which can then be transfered to the printer and upon printer setup can be built. Next, the printed part must be removed to be prepared for application through certain post-processing steps, which directly depend on AM technology and materials used.

In the remainder of this section, we will first review different AM technologies and discuss various topics related to design for AM.

13.1 Additive Manufacturing Technologies

There is a variety of AM processes which differ in material and deposition process. Generally, they can be classified as:
1. Material Extrusion
2. Powder Bed Fusion
3. Direct Energy Deposition
4. Vat Photo Polymerization
5. Material Jetting
6. Binder Jetting
7. Sheet Lamination

> This section provides a brief overview on these AM technologies, however for more detailed discussion and analysis, the readers are referred to :
> - *Additive Manufacturing Technologies*, Ian Gibson, David Rosen, Brent Stucker, Springer, 2015.

13.1.1 Material Extrusion

Material extrusion or Fused-Deposition Modeling (FDM) is an AM process through which molten filaments are extruded from a nozzle and deposited on a substrate to build the part layer by layer. Most common materials in FDM are thermoplastics such as ABS or PLA, however other materials such as composite fibers, concrete, or filaments with metal infills are also used.

FDM is fairly robust with respect to build scale and material. This, along with other advantages such as ease of use, portability, affordability, and safety make FDM very promising in producing functional parts in applications such as: There are however certain challenges associated with FDM. Since the FDM parts are composed of molten filaments of material stacked on top of each other, the interlayer adhesion (or lack thereof) plays a critical role in how the part behaves under structural loading. As a general rule, the *smaller* the print area,

13.1 Additive Manufacturing Technologies

(a) Large-scale printing
(http://mashable.com)

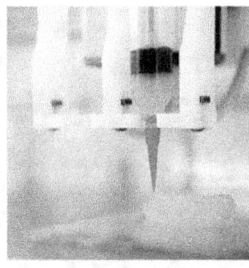
(b) Biomedical customized parts
(http://news.cornell.edu)

(c) Embedded electronics
(http://www.voxel8.com)

(d) Printing in space
(http://madeinspace.us)

Figure 13.2: Some of the FDM application.

the *better* the interlayer fusion. If the print area (at each layer) is sufficiently small, it has not fully solidified and *sticks* better to the material being deposited. On the other hand, if the print area is too large, the molten material is deposited on a cold solidified surface which 1) lacks proper adhesion resulting in anisotropy along build direction and 2) causes warpage. Since this anisotropy is *process induced* as opposed to *intrinsic* (e.g. composite materials), it is usually unfavorable and must be avoided. Standardization of process parameters and material properties is currently an active area of research. Other limitations of FDM the need for extraneous support structure and coarse resolution which makes it not suitable for printing fine details and can result in poor surface quality.

13.1.2 Powder Bed Fusion

Powder Bed Fusion (PBF) is another class of AM technologies, in which a laser beam selectively scans regions of the powder bed to build the part through particle fusion. Based on the level of fusion, there are essentially two types of PBF processing, 1) *sintering*, where the particles are partially fused together and 2) *melting*, where the heat source is strong enough to melt the powder and create a melt pool as it scans the surface of the bed.

Selective laser sintering (SLS) is a popular PBF technology for building plastic parts. Recently, a multi-jet fusion (MJF) is developed that fabricates plastic parts by fusing particles through adhesive agents and curing. In SLS, there is no need for support structures and the mechanical properties are usually better than FDM, while the surface finish can be too coarse for some applications. However, perhaps the most important limitation is due to material cost and the fact that unused powder can become degraded and wasted. It is also possible that powder is more dense at the bottom of the reservoir, which can induce anisotropy along build direction. Direct Metal Laser Sintering (DMLS) is specific to metal alloys. Additionally, Selective Laser Melting (SLM or LM) is used to fabricated strong and dense metallic or ceramic parts, however compared to DMLS there are less metals available for this process. Electron Beam Melting (EBM) is another PBF process for metals which compared to SLM builds the part faster while at the price of reducing surface quality.

Industrial interest in Metal based PBF processes has significantly increased in the recent years as more standard metal powders are becoming available. However, in SLM and DMLS the metal parts might require support structure to reduce thermal residual stresses. Also, the whole process from printer to powder can become quite costly. Compared to SLM, EBM parts are built faster and require less support structure.

13.1.3 Direct Energy Deposition

Directed Energy Deposition (DED) process is an another AM technology that uses metal powders to fabricate 3D parts. Unlike PBF, in DED the metal powder is melted as it is being deposited. Laser Engineered Net Shaping machine (LENS^{TM}) is also a closely related variation of DED.

The process can use wire or powder of mostly metals and can be extremely useful in engineering novel alloys however, currently there are not many materials available for this process.

(a)

(b)

Figure 13.3: (a) LENS^{TM} process and (b) final printed part after machining.

13.1.4 Vat Photo Polymerization

Vat photo polymerization or StereoLithogrophy Apparatus(SLA) is the oldest AM process where a laser beam scans a vat of UV-curable liquid photopolymer to create the part through layer by layer polymerization.

SLA parts have great accuracy and surface finish. The build process can be quite efficient as well, however, the post-processing stages of removing resin and curing are typically quite time intensive and laborious.

Further, the number of photo resins available are quite limited and the SLA parts are brittle and often require support structure.

13.1.5 Material Jetting

In Material Jetting (MJ), droplets of wax-like polymers and plastic materials are selectively deposited.

Material jetting parts typically have good surface quality and accuracy, which make this process very well-suited for printing casting molds. Also, MJ is perhaps the most popular

process for printing multi-material and multi-colored parts.
One of the challenges associated with material jetting is the limited number of wax-like materials.
The build process is typically slow due to material droplet deposition and subsequent milling operations. Further, the MJ parts are brittle and fragile.

13.1.6 Binder Jetting

Binder Jetting (BJ) can be used for a wide range of materials, where liquid adhesive material is selectively deposited to bond powders.
Binder jetting is fast and cheap, and can essentially use any material that is available in powder form to create multi-material and multi-color parts. Apart from prototyping, it can be used for casting molds and cores.
However, without proper post-processing, the parts are typically fragile with poor mechanical properties.

13.1.7 Sheet Lamination

In sheet lamination process or laminated object manufacturing (LOM), sheets of material are bonded together with adhesive material, while the required shape of each layer is created by laser-cutting or knife.
For this process, any sheet material that can be rolled such as paper, plastic, and some metals can be used.
LOM is relatively fast and cheap. It is also possible to fabricate larger parts since it involves no chemical reaction. However the number of materials can be limited and strength along build direction heavily depends on the adhesive material. Further, the process is not very well-suited for complex geometries and can create lots of material waste. Also, to increase the surface finish quality, significant post-processing might be needed.

13.2 Geometry Representation

STereoLithography (most commonly known as STL) format, is a boundary representation of the three dimensional object and is widely used in industry for data transmission, especially in additive manufacturing. In fact, STL was invented in 1987 by Albert Consulting Group for 3D Systems.
In essence, STL format approximates the surface of a geometry with a set of connected flat triangles as shown in Figure 13.4.

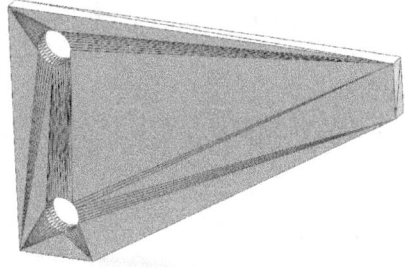

Figure 13.4: STL file of the support bracket.

STL files can be stored in two formats, ASCII and Binary. The former contains printable characters while the latter contains encrypted data to reduce file size.

13.2.1 ASCII format

In ASCII format, the file begins with *solid* (all lowercase) and ends with *endsolid*. In between, the file specifies each of the triangles through *facet normal* (outward) and positions of the three vertices.

The order of vertices must be according to right-hand rule, so that it is consistent with the facet normal. The vertices are wrapped with the keywords *outer loop* and *endloop*. Description of each vertex begins with the keyword *vertex* which is followed by coordinates in X, Y, and Z directions.

The following shows the first triangle for the support bracket of Figure 13.4.

```
solid
  .
  .
  .
  facet normal  -1.00    0.00           0.00
  outer loop
  vertex    0.00    7.624835e-003    3.000000e-003
  vertex    0.00    5.000000e-002    3.000000e-003
  vertex    0.00    7.624835e-003    0.00
  endloop
  endfacet
  .
  .
  .
endsolid
```

13.2.2 Binary format

For large and complex geometries, the ASCII files can become extremely large and challenging to work with. To overcome this issue, the STL files can be stored in binary format, which contain the following data:

- An 80-character header (UINT8[80])
- The total number of triangles which is an unsigned integer with 4 bytes (UINT32)
- The descriptions of each triangle through 12 floating-point numbers (each 32 bits), similar to ASCII format:
    1. Facet normal components (REAL32[3])
    2. X, Y, Z of vertex 1 (REAL32[3])
    3. X, Y, Z of vertex 2 (REAL32[3])
    4. X, Y, Z of vertex 3 (REAL32[3])
- Attribute byte count (UINT16)
- end

(R) For visualizing or manipulating STL files, readers can use MeshLab^{TM} (http:

//www.meshlab.net), which is a free, open-source, and cross-platform software.

13.2.3 Limitations of STL and Other Options

As was mentioned, STL is the standard format for many applications and is extremely popular. However, there for modern applications, it can pose many challenges due to following limitations:

1. STL files do not provide information on units, i.e. it is not possible to distinguish whether the part is in meters or milimeters.
2. It is not possible to assign texture or graded materials to an STL file.
3. Complex industrial models and hierarchical designs require millions of triangular faces to capture all the features. This would drastically increase the size of STL files.

Due to these limitations there is a great need for more modern formats that are capable of capturing geometric features more accurately and represent material and textural attributes. In 2011, American Society for Testing and Materials (ASTM) proposed the additive manufacturing format (AMF) to address the issues associated with STL. AMF is non-proprietary, allows curved triangular faces, can represent material gradients and textures, and is backward compatible with STL.

Another replacement candidate for STL is 3MF format developed by Microsoft. Similar to AMF, 3MF is capable of handling colors, materials, textures, and supports traditional STL files. Further, 3MF can be used for different subtractive manufacturing processes such as CNC, and is seamlessly integrated within Windows.

Consider Figure 13.5, which is a highly complex model of a motherboard.

Figure 13.5: A highly complex model.

Figure 13.6 illustrates the differences in file size or the STL, AMF, and 3MF formats. Observe that using ASCII STL files are highly inefficient and should be avoided for complex models. Although binary STL files are a bit smaller than ASCII, they are still far from ideal.

On the other hand, both AMF and 3MF file, if encoded properly, are comparable and quite efficient.

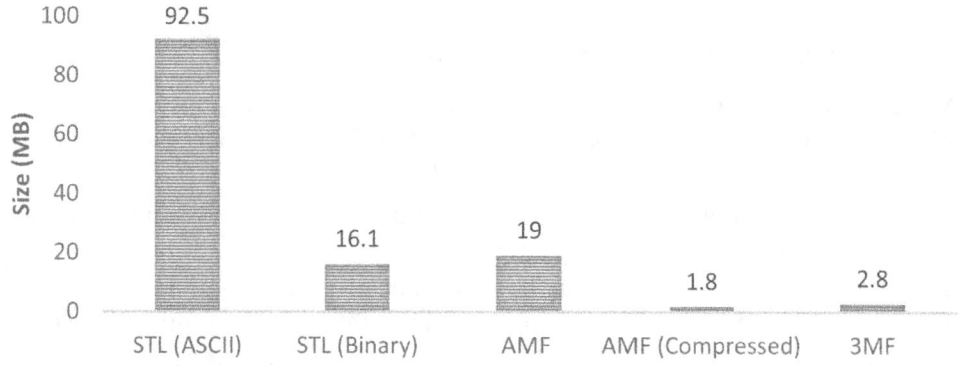

Figure 13.6: Size comparison between STL, AMF, and 3MF.

13.3 Design for Additive Manufacturing

As additive manufacturing technologies are transforming from solely prototyping to actual functional parts, the quality of print to produce defect-free parts becomes more and more crucial. However, each AM process has its benefits and challenges which need to be considered at the design stage.

Over the years, designers and researchers have produced general design rules based on experience and practical knowledge. In this section, we will first discuss some of the issues associated with different AM technologies and explain the design guidelines. Finally, we will consider integrating these AM related constraints with our topology optimization framework.

® Following references offer more details on design guidelines for additive manufacturing:

- *Functional Design for 3D Printing : Designing 3D printed things for everyday use*, C. T. Smyth, 2nd edition, 2015.

13.3.1 Support Structure

Support structure generation in AM is based on the overhang concept which states that *if the angle between the boundary normal and the build direction exceeds a certain threshold, then support structures are needed at that point*. For instance, for the design and the build-direction illustrated in Figure 13.7a, the subtended angle α is illustrated in Figure 13.7b. Given a threshold $\hat{\alpha}$ (typically around 135°), boundary points with $\alpha > \hat{\alpha}$ are considered overhanging, and require support as illustrated in Figure 13.7c. The union of all such support structures results in a support volume as illustrated in Figure 13.7d. The fill-ratio, i.e., material density, of support structures is typically less than that of the primary design.

13.3 Design for Additive Manufacturing

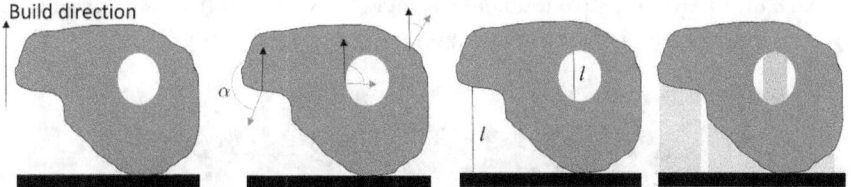

Figure 13.7: (a) Build-direction. (b) Subtended angle. (c) Support length. (d) Support volume.

There are a number of AM processes that may require support structures yet for different reasons. In FDM, support structures as the plastic part is built, each layer must be properly supported by previous layers due to weight. For overhang surfaces, this support is not provided by design itself, therefore extra support structures must be printed so that the molten material would not droop. Figure 13.8 demonstrates Failure of FDM parts due to lack of support.

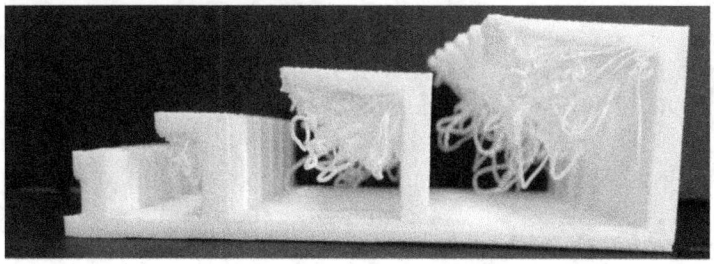

Figure 13.8: Failed FDM prints in absence of support.
(https://innovationstation.utexas.edu/tip-design)

Unlike FDM, in Powder Bed Fusion (PBF) processes, the weight of part is sustained by the bed of powder, however some metal PBF technologies, such as Selective Laser Melting (SLM), Direct Metal Laser Sintering (DMLS), or Electron Beam Melting (EBM) may require support structures to diffuse the accumulated heat in overhang surfaces. In other words, Support structures in metal additive manufacturing work as a heat sinks to dissipate excessive heat into the surrounding powder. Figure 13.9 shows an optimized GE engine bracket printed by Arcam (EBM) metal 3D printer.

Figure 13.9: Optimized GE engine bracket printed by Arcam (EBM) metal 3D printer.

If proper support is not provided, the temperature difference can result in dramatic distortions, warpage, curling, or burning; thus failure of the print job:

Figure 13.10: Failed metal parts in absence of support.
(Mertens, Clijsters, Kempen, Kruth, *Optimization of Scan Strategies in Selective Laser Melting of Aluminum Parts With Downfacing Areas*, ASME. J. Manuf. Sci. Eng. 2014.)

Printing support structures directly adds to the build-time and material cost. Material costs can be substantial in AM; for example, the largest percentage cost for metal AM, besides the machine cost that is amortized, is material cost (18%). Further, support structures can be hard to remove (and sometimes even inaccessible), leading to the post-fabrication (clean-up) cost. Post-fabrication costs make-up for about 8% of AM product cost.
Thus, clearly it is favorable to reduce the amount of support structure as much as possible. This can be achieved through various strategies:
1. Optimal build orientation
2. Optimal support structures
3. Design paradigms

We will briefly review each of these strategies.

Optimal build orientation

The most popular approach to reducing the amount of support structure is to find an optimal build orientation which produces least amount of support structure.

For the majority of 3D printers available, there is only a single build direction which is set *a priori*. Typically, support structure is not the only deciding factor in choosing a build direction and other parameters such as surface finish, flatness and cylindrical errors, build time, and mechanical strength are also considered. Numerous methods have been developed over the years, which are mostly based on optimizing with respect to the weighted average of multiple parameters p_i, where the following optimization problem is solved:

$$\underset{p_i}{minimize} f = \sum w_i p_i \\ \text{subject to} \\ \sum w_i = 1 \qquad (13.1)$$

As was previously discussed, although standard gradient-based methods are extremely popular in many applications, they are not suitable for solving 13.1, since the problem is non-differentiable and highly nonlinear with respect to support structure. Thus, most of the

13.3 Design for Additive Manufacturing

methods proposed for solving this class of problems are based on non-gradient optimization methods, such as genetic algorithms, which involve random selection and modification of solutions rather than deterministic convergence.

With the advent of off-axis printing, it is also possible to print each section of the design at a different orientation to avoid overhang angles.

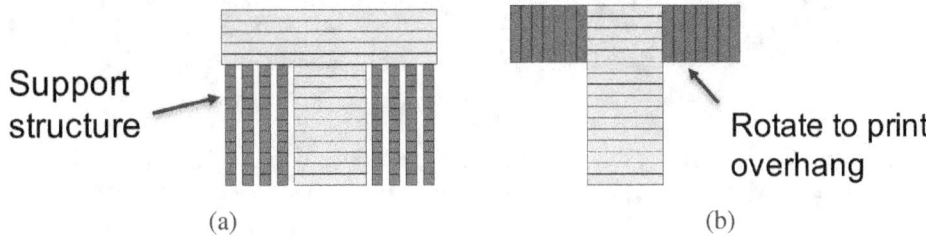

Figure 13.11: Using multiple build directions to reduce support volume.

Efficient support structures

Although optimizing the build direction can significantly reduce the amount of support structure, it may not necessarily eliminate it. Thus, we also need to generate more efficient supports for these inevitable cases. Given the build direction, the easiest and most straightforward way to generate support structures is by printing vertical support beams, however this may not be the most efficient method.

One very popular approach in generating more cost-effective supports is to use a different material for printing supports. These materials can be either cheaper which directly reduces the cost or soluble which makes it easier to remove and less laborious. While this approach is very well-suited for fused deposition modeling, some of the other AM technologies such as powder-bed fusion or vat photo-polymerization techniques are not capable of printing with multiple materials.

Figure 13.12: Soluble support structures.

Another approach is to modify the structure of support structures. One type of structures that are used for creating more efficient support structures is the so-called *tree-like* structures:

(a) Vertical support. (b) Tree-like support.

Figure 13.13: Tree-like structures to reduce support volume.
(Vanek, Galicia, Benes, *Clever Support: Efficient Support Structure Generation for Digital Fabrication*.)

It is well-known that in some AM processes such as FDM, small overhang surfaces with horizontal distance of about 5mm to 20mm can be fabricated without the need for additional support. Thus, yet another strategy for creating more efficient support structures is through generating scaffolding structures:

 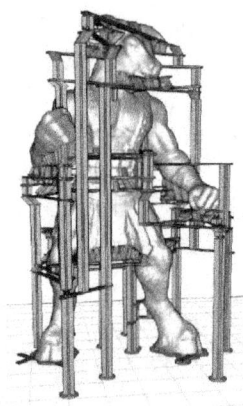

(a) Tree-like support. (b) Scaffolding support.

Figure 13.14: Scaffolding structures to reduce support volume.
(Dumas, Hergel, Lefebvre, *Bridging the Gap: Automated Steady Scaffoldings for 3D Printing*)

> (R) Meshmixer^{TM} (www.meshmixer.com) is an open-source software that can create these tree-like support structures.

Although the methods discussed above can potentially be used for different AM technologies, they are mostly well-suited for FDM and polymers. As was previously mentioned, for metal AM, support structures are needed to dissipate excessive heat and are much harder to remove. On strategy for creating efficient supports for metal AM is supports that are not fully attached to the print job while successfully reduce residual stresses. Figure 13.15 (left) shows the warpage occurred during EBAM process in the absence of support structures. Figure 13.15 (right) demonstrates the effectiveness of extraneous support, as the warpage is clearly mitigated.

13.3 Design for Additive Manufacturing

Figure 13.15: Warpage in EBAM printed part in the absence of support structures (left) and mitigation of support in presence of support (right) contact.
(Cooper, Steele, Cheng, Chou, *Contact Free Support Structures for Part Overhangs in Powder-Bed Metal Additive Manufacturing*)

Design paradigms

Yet another approach to reducing support structure is by *modifying the design* in such a manner that it would *require less support* when it is being printed. Various types of modification have been proposed, some of which will be discussed in this section.

Decomposing. One approach to altering the design for print is decomposing it to self-supported components and printing each component individually.

Lattice structures. An effective strategies to improve AM designs is leveraging AM capabilities in fabricating highly complex shapes to create hierarchical lattice structures; where each unit cell is self-supported.
However, there is a caveat to blindly using this approach, especially for functional parts. Since large number of stress concentrations are introduced, failure might occur due to fatigue or buckling. On the other hand, if physics of the problem is properly considered, hierarchical designs can generate light-weight high-performance structures which can also exhibit auxetic structural properties such as negative Poisson's ratio.

Origami and 4D printing. Another approach to reducing support structure is exploiting origami algorithms, where a complex geometry is decomposed into many polygons that can be unfolded onto a 2D surface and printed without the need for support structure. Once the 2D part is printed, it is folded such that the original design is created.
It is also possible to create designs which dependent on certain environmental factors, can transform from one shape to another given sufficient time. This technology is called 4D printing since the configuration of the design depends on 3 spatial dimensions and 1 temporal dimension. For instance it is possible to print complex shapes in 2D, where no support structure is needed, and by changing for instance the temperature it would self-assemble to the intended 3D object.

13.3.2 Anisotropy

Material anisotropy and weakness along build direction, especially in FDM, is an important issue which becomes more critical when the part is functional and must satisfy strength-related constraint. There are mainly two types of anisotropy, namely:
1. Intrinsic
2. Process-induced

Composites are an example of materials that exhibit intrinsic anisotropy, where directional preferences in mechanical and thermal behaviors are *intentionally* created to enhance performance. On the other hand, process-induced anisotropy, which occurs in many AM processes, is the result of process limitations and is often unfavorable since it introduces uncertainty in material properties.

The material anisotropy in FDM technology is due to lack of interlayer fusion between layers. In other words, the adhesion between layers is limited by the contact surface of filaments of material as they are being deposited, as illustrated below:

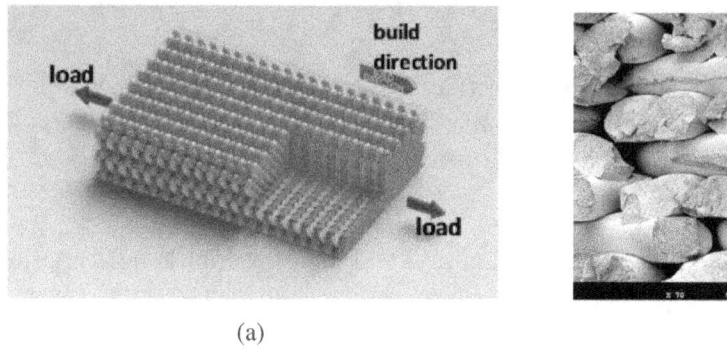

Figure 13.16: Anisotropy in FDM.
(Riddick, Haile, Wahlde, Cole, Bamiduro, Johnson, 2016. *Fractographic analysis of tensile failure of acrylonitrile-butadiene-styrene fabricated by fused deposition modeling*, Additive Manufacturing)

Material anisotropy can also occur in metal processes, where the thermal history (heating and cooling) at every point might create porous microstrucutres throughout the print job, as shown here:

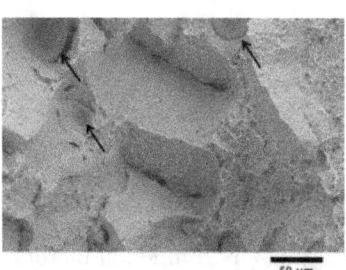 

(a) Porosity in SLM part.   (b) Fracture surface at porous region.

Figure 13.17: Anisotropy in SLM.
(Simonelli, Tse, Tuck, *Effect of the build orientation on the mechanical properties and fracture modes of SLM Ti–6Al–4V*, Materials Science and Engineering.)

13.3.3 Surface quality

Based on the resolution of the printer and layer thickness, quality of surface finish can vary from one AM process to another.
Consider Figure 13.18 for FDM parts, the surface quality drastically decreases (or surface roughness increases) for $70° \leq \theta \leq 90°$ due to stair stepping effect:

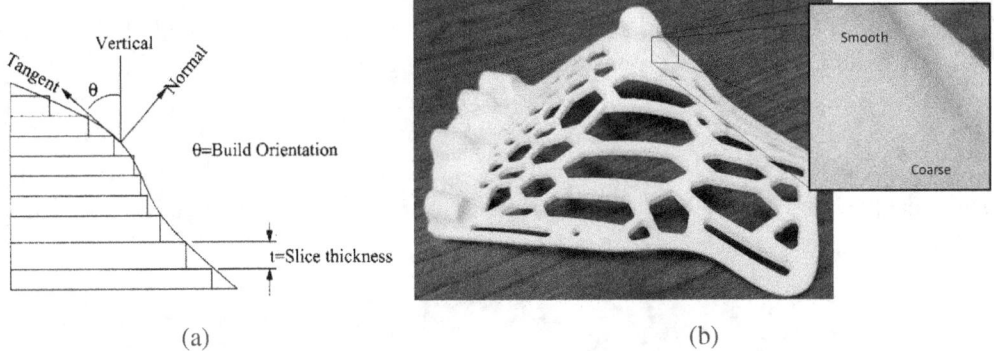

Figure 13.18: Surface quality in FDM.

For metal processes, the surface quality is directly related to particle size, laser power, and deposition process. Figure 13.19 shows the surface quality in the two metal-based AM process.
For instance, The average particle size for selective laser melting (SLM) process is generally lower than that of electron beam melting (EBM) ($36 \mu m$ for EOS Ti64 powder versus $60 \mu m$ for Arcam Ti64 powder). Typically, SLM parts have better surface quality and fatigue limit, while EBM parts have higher ductility.

(a) SLM surface quality

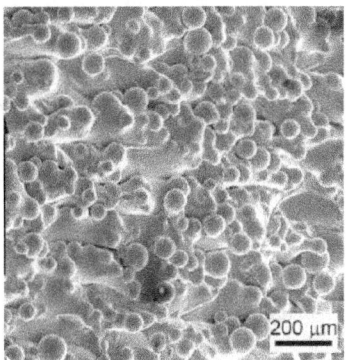
(b) EBM surface quality

Figure 13.19: Surface quality in SLM and EBM.
(Rafi, Karthik, Gong, Starr,Stucker, *Microstructures and Mechanical Properties of Ti6Al4V Parts Fabricated by Selective Laser Melting and Electron Beam Melting*, J. of Materi Eng and Perform)

13.3.4 Shrinkage and warping

One of the major issues in AM is that as the part is being printed, some areas of the print job are heating up while others are cooling down and at different rates. As the temperature gradient grows higher, some portions of the part might shrink and warp; resulting in failure of the print.

In FDM and polymers, it is known that the material contracts as it is cooling down. Of course some polymers such as Acrylonitrile Butadiene Styrene (ABS) have higher thermal expansion coefficient than other polymers such as polylactic acid (PLA). For ABS parts (less so for PLA parts), if the adhesion between layers is not strong enough, the shrinkage can cause delamination. Thus, to ensure better bonding, we need the material underneath to remain hot when new material is deposited. To this end, 1) the build chamber temperature must be around $70°C$-$90°C$ and 2) it is preferable that the print area at each cross-section remains as small as possible. Also, to prevent the first layer to shrink and warp, we must 1) pre-heat the build plate and 2) use adhesives so the material properly sticks to the substrate.

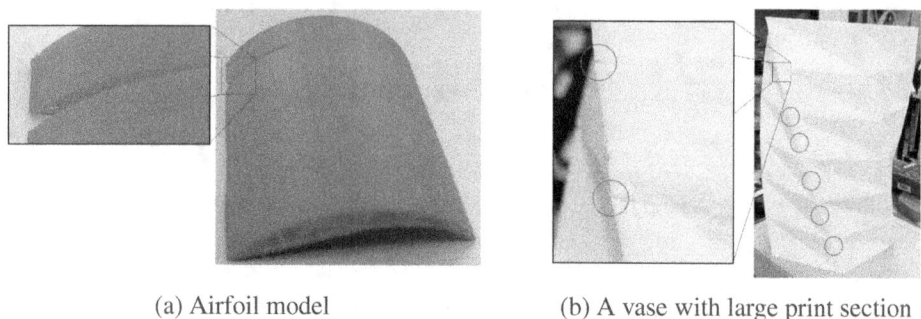

(a) Airfoil model   (b) A vase with large print section

Figure 13.20: Warping in FDM.

In metal AM, the rates of melting and solidification are extremely high. The high solidification rate introduces a great amount of thermal stresses which can cause warping in the part. Warping in metal based processes such as SLM, DMLS, or LENS poses significantly more challenging compared to FDM process for polymers, since the temperatures are much higher, heat source is more localized, failure in print is much more costly, both in terms of time and material. Figure 13.21 demonstrated the warped build plate used for LENS. Powder temperature, laser power, laser scanning pattern, and support structure are among the most important parameters in preventing warpage in metal AM. Also, once the part is built, it must usually be heat treated so the residual stresses are relieved.

Figure 13.21: Warping in metal AM (LENS).

13.3.5 Infill patterns

In many cases, to reduce weight of part, material usage, and build time the interior of AM parts are not fully solid. Instead, the interior structure can comprise different types of infill; some of the most commonly used structures are:

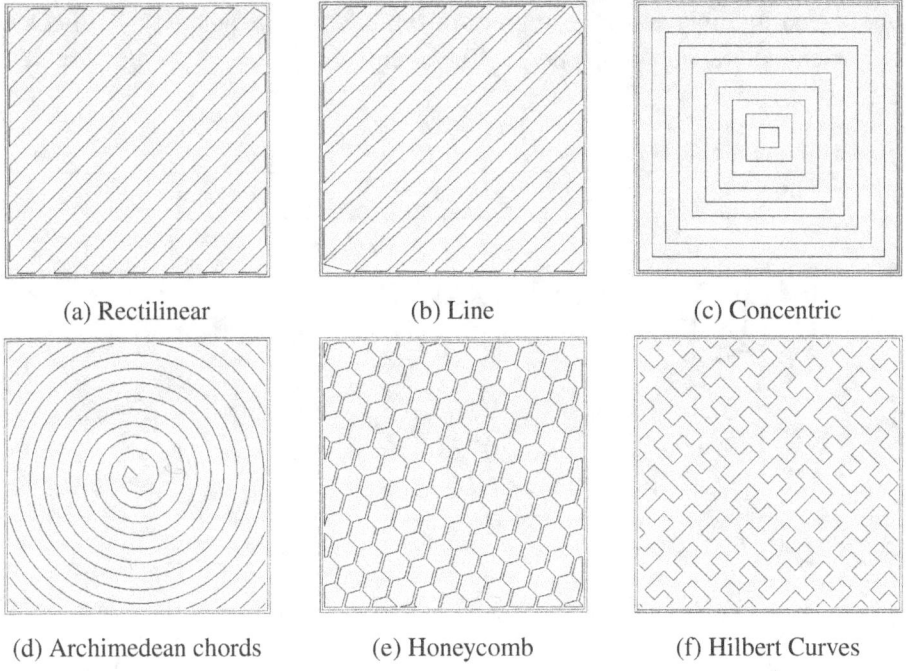

(a) Rectilinear (b) Line (c) Concentric

(d) Archimedean chords (e) Honeycomb (f) Hilbert Curves

Figure 13.22: Different infill patterns.

Among these, rectilinear and honeycomb structures are widely used. There is no particular advantage in using rectilinear patterns, honeycomb on the other hand exhibit great strength for the amount of material used and are exploited in various applications.

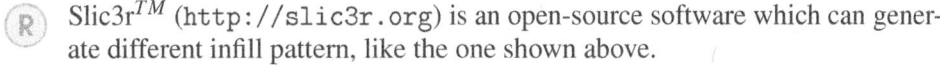 Slic3r^{TM} (http://slic3r.org) is an open-source software which can generate different infill pattern, like the one shown above.

13.4 Lattice Structures

As was mentioned in the previous chapters, additive manufacturing processes, unlike traditional subtractive technologies, are relatively insensitive to geometric complexities of design. However, fabricating parts through AM can become expensive due to high material cost, build time, and post-fabrication cost. Therefore, it is crucial to lighten the design. A class of such methods, besides topology optimization, to lighten a design is cellular lattice structure design, where a fully-dense design is made porous through introduction of periodic cellular structures in the interior of the design.

Typically, lattice designs are used in lieu of topology optimization when either (1) there is no significant load (for example, lightening of a fancy pencil holder), or (2) performance constraints such as stress and buckling constraints are not critical. In this chapter we will consider creating cellular lattice structures from given a solid model.

13.4.1 Lattice Unit Cell

There are different types of unit cells that can be used in generating lattice structures; these are summarized in Table 13.1. The choice of the unit cell is dictated by various factors including manufacturability, aesthetics, stiffness requirements, etc.

Table 13.1: Lattice unit cell types

Unit Cell	Name	#Nodes	# Edges
	Box	8	12
	Body Centered Cubic (BCC)	9	8
	Box + BCC (BOXBCC)	9	16
	Body Centered Cubic + Z edges (BCCz)	9	12
	Face Centered Cubic (FCC)	14	24
	Face Centered Cubic + Z edges (FCCz)	14	28

continued ...

13.4 Lattice Structures

Unit Cell	Name	#Nodes	# Edges
...continued			
(octet lattice image)	Octet	14	36

13.4.2 Example

■ **Example 13.1 Circular plate with lattice structure**

1. Load the *CircularPlateHoleProject*.
2. Under *Lattice Design* menu:
3. Set type to *BOX*.
4. Set unit size to 10 mm.
5. Set number of divisions per edge to 20..
6. Set the fill ratio to 0.10
7. Keep critical surfaces; this means that all the load-bearing and restrained faces will be retained as solid material.
8. Set number of smoothing iterations to 2.
9. Create the lattice design.
10. Figure 13.23 shows the lattice structure that is created.

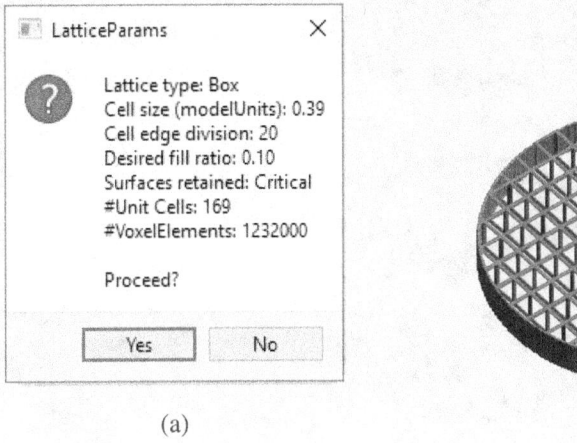

Figure 13.23: Circular plate with hole (a) lattice parameters and (b) lattice structure.

11. One can repeat the lattice design for other cell types as illustrated below.

Table 13.2: Circular plate with different lattice unit cell types

Lattice Design	Unit Cell
	Box
	BCC
	Box-BCC
	BCCz

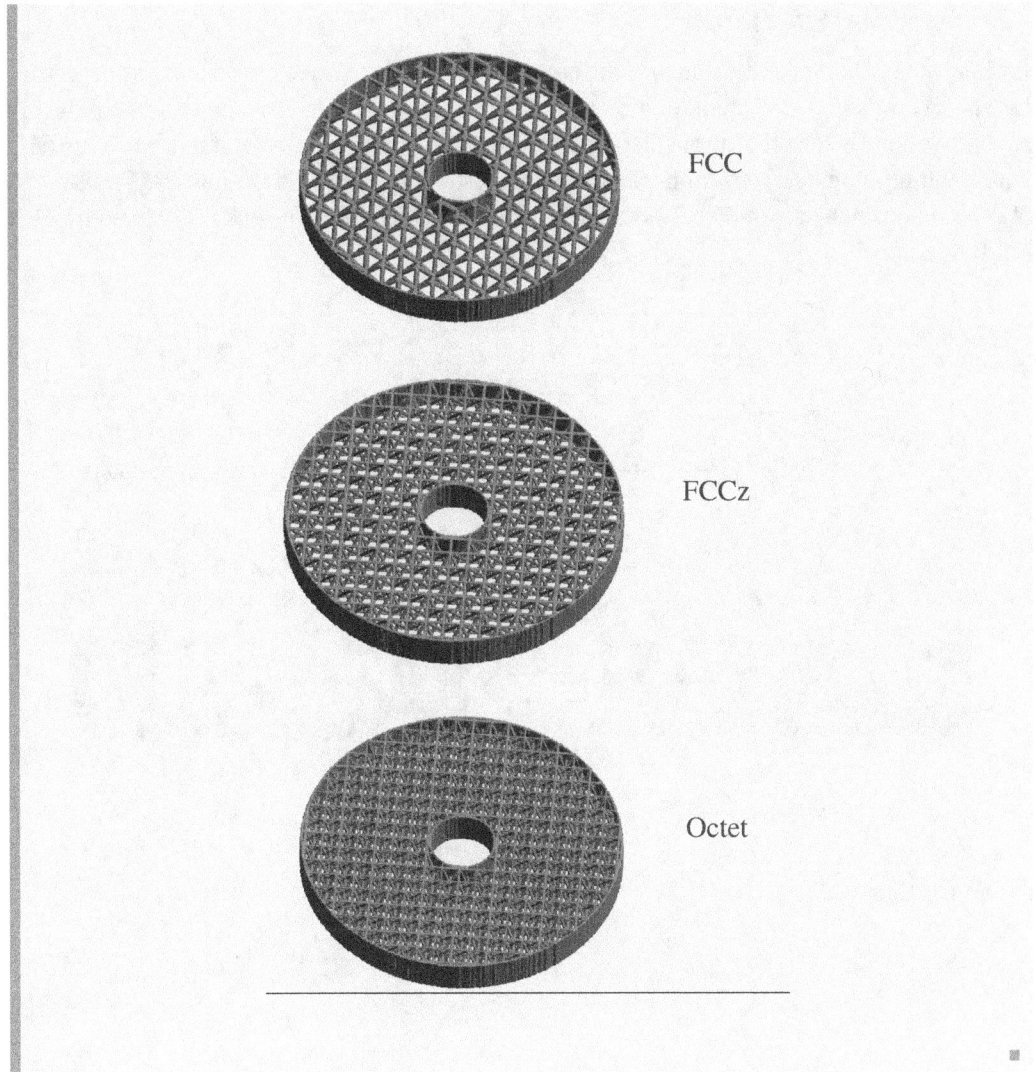

13.5 Topology Optimization for Additive Manufacturing

Additive Manufacturing and topology optimization complement each other in that organic and complex designs generated through TO can be manufactured through AM technologies. On the other hand, the cost of AM parts increase significantly with material usage. Thus, optimizing designs can be crucial in saving material usage, build time, and post-fabrication time. However, there are certain challenges in TO for AM which need to be addressed before the two fields can be seamlessly integrated.

The focus of this section is primarily on integrating TO and AM through 1) consideration of manufacturing constraints imposed by AM processes and 2) exploiting the capabilities of AM in free-form fabrication to explore more complex designs and improve performance. To this end, we will discuss topology optimization considering support structure.

13.5.1 Support Structure

Earlier in this chapter, we discussed different techniques to reduce amount of support structure needed in some AM technologies such as FDM or SLM. Another approach to reducing support volume is to include consideration for support structure as a manufacturing constraint within our topology optimization framework so that the optimized design would require less (or no) support when printed. Generally, this can be achieved in various ways including but not limited to:

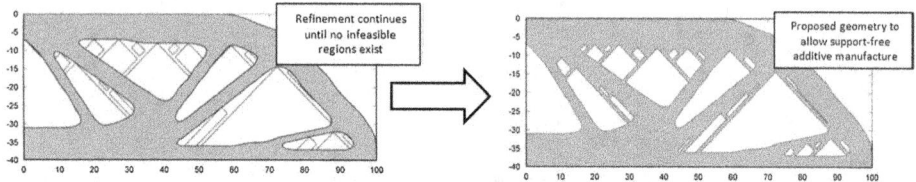

(a) Post-optimization
(Leary,Merli,Torti,Mazur,Brandt, *Optimal topology for additive manufacture: A method for enabling additive manufacture of support-free optimal structures*, Materials & Design)

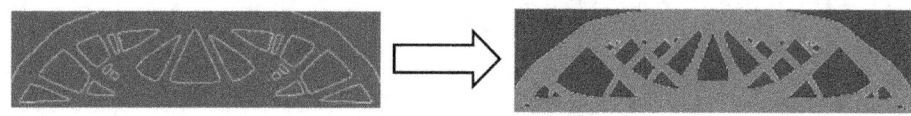

(b) Filter sensitivity
(Gaynor,Meisel,Williams,Guest, *Topology Optimization for Additive Manufacturing: Considering Maximum Overhang Constraint*, 15th AIAA/ISSMO Multidisciplinary Analysis and Optimization Conference.)

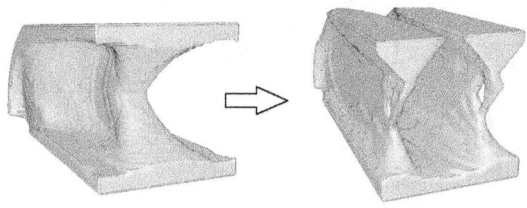

(c) Filter sensitivity
(Langelaar, *Topology optimization of 3D self-supporting structures for additive manufacturing*, Additive Manufacturing.)

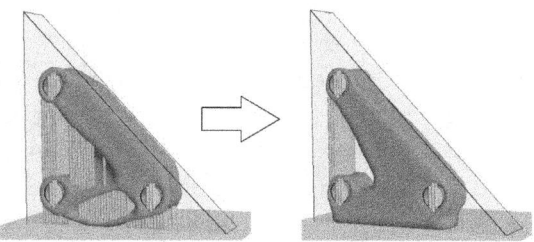

(d) Augment support sensitivity
(Mirzendehdel, Suresh, *Support structure constrained topology optimization for additive manufacturing*, Computer-Aided Design.)

Figure 13.24: Support constraint in topology optimization.

13.5 Topology Optimization for Additive Manufacturing

1. Post-optimization: Modify the design after the optimization is finished to create a self-supporting design.
2. Filter sensitivity: At every optimization step, apply a filtering scheme to the underlying sensitivity field to consider overhang surfaces.
3. Augmented support sensitivity: At every optimization step, compute a support structure sensitivity and augment it to the underlying sensitivity field.

Figure 13.24 illustrates an example of each methodology.

Most of the approaches towards aim to exploit the fact that if one could *eliminate all overhanging surfaces*, then support structures can also be eliminated. But, this may not be an effective optimization strategy for the following reasons:

1. *Eliminating all overhanging surfaces may not be possible.* It has been demonstrated that one can eliminate overhang surfaces in certain 2D problems (Figures 13.24a to c). However this is unlikely to be successful in general, especially in 3D (as the numerical examples in Section 4 demonstrate). As was also suggested in [13.3], "...there will probably be instances where it is not necessary for all support structure to be eliminated and so the user should be able to have some control over the strength of the penalty function.".
2. *The overhang constraint does not penalize support volume.* Two overhanging surfaces with equal subtended angle will be penalized equally, although the support volume associated with one may be much larger than the other. To avoid such contradictions, a direct constraint on the support volume is desirable.
3. *Penalizing just the overhanging surfaces is insufficient.* Support volume may be enclosed between an overhanging surface and an opposing surface, as illustrated in Figure 13.25. To reduce support volume, both surfaces must be penalized, for example, by moving them closer to each other as illustrated. By penalizing the overhanging surface, only half the problem is addressed.

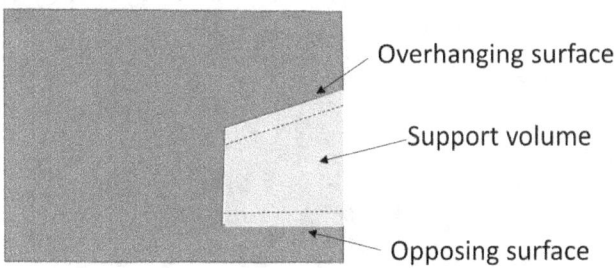

Figure 13.25: Moving either the overhanging or its *opposing* surface changes the support volume.

4. *Self-supported sharp corners might be sub-optimal under stress or fatigue.* When considering stiffness, creating self-supported designs similar to (a) and (b) in Figure 13.24 may be optimal; however when more realistic objectives and constraints are considered, the introduced sharp corners could become quite problematic by stress concentration and crack initiation.

In this section we will discuss the augmented support sensitivity approach which is based on tracing the Pareto frontier and the formulation relies on two steps:

1. Dynamically estimating the support volume as the topology evolves
2. Imposing constraints on the support volume through topological sensitivity methods.

Consider the first step of dynamically estimating the support volume. In this framework, we assume that support structures are vertical. Therefore, the support volume is simply the integral of the support length over the boundary, multiplied by a suitable fill-ratio, (see Figure 13.7d), i.e.

$$S = \gamma \int_{\alpha \geq \hat{\alpha}} l_p d\Gamma \tag{13.2}$$

where:

$$\begin{aligned} &S: \text{Support strucuture volume} \\ &\alpha: \text{Subtended angle} \\ &l_p: \text{Lenght of support structure at boundary point p} \\ &\gamma: \text{Fill ratio (relative material density) of support structures)} \end{aligned} \tag{13.3}$$

In Equation 13.2, the exact value of the fill ratio is not critical; it can be assumed to be 0.5, without a loss in generality.

Further, for short overhangs, it is well known that support structures are not needed. For example, for FDM, the allowable overhang [13.3] can be approximated via:

$$h(mm) = \begin{cases} 5 + 40(1 - \dfrac{\alpha}{\pi}) & \dfrac{3\pi}{4} < \alpha \leq \pi \\ \infty & 0 \leq \alpha \leq \dfrac{3\pi}{4} \end{cases} \tag{13.4}$$

Thus, at any point on the boundary, if the subtended angle is α, support structures are not needed if the overhang distance is less than h given by Equation 13.4. In implementation, we search for self-supporting boundary within a distance given by Equation 13.4; see Figure 13.26.

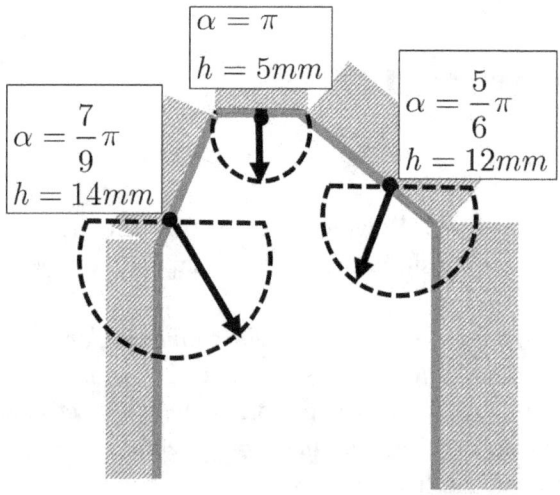

Figure 13.26: Searching for self-supporting boundary.

13.5 Topology Optimization for Additive Manufacturing

Options for imposing support constraint

Consider the three-hole bracket of Figure 13.27 where the two left side holes are fixed and the right-hand side hole is subject to a downward unit load. The underlying material is assumed to be isotropic ABS plastic with Young's modulus of $E = 2GPa$ and Poisson ratio of $v = 0.39$.

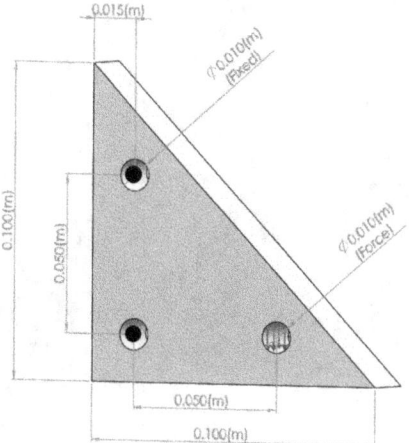

Figure 13.27: Three-hole bracket geometry and boundary condition.

Next, we want to reduce volume of the initial design by 50% while retaining its stiffness by solving the following compliance minimization problems:

$$\underset{\Omega \subset D}{minimize} \{|\Omega|, J\}$$
$$\text{subject to}$$
$$\mathbf{Ku} = \mathbf{f}$$
(13.5)

Figure 13.28 illustrates the progression of the optimization process in Pareto up to a volume fraction of 0.5. Observe that optimization begins with a volume fraction of 1.0, and generates multiple topologies that lie on the Pareto front. This will play an important role in the proposed method for constraining the support structure volume. Further, we do not rely on a velocity field concept to move the boundary; instead, we use fixed-point iteration to converge to Pareto-optimal designs.

Next consider the challenge of imposing support volume constraint. Perhaps the simplest strategy is to impose an absolute constraint as in:

$$S \leq S_{max} \tag{13.6}$$

However, this places an unreasonable burden on the designer to arrive at an absolute value for the upper limit a priori. Instead, we consider relative upper bound constraints. Specifically, recall that in the Pareto method, one generates multiple topologies for various volume fractions, i.e., one can solve the unconstrained problem, and store reference support volumes $\eta S_{unc.}(v)$ at intermediate volume fractions. For example, Figure 13.29 illustrates the support volume $\eta S_{unc.}(v)$ for the unconstrained problem. The support volume curve is, in general, non-smooth, unlike the compliance curve in Figure 13.28. Next we impose a relative constraint with respect to $\eta S_{unc.}(v)$, via a user-defined parameter $\eta (0 < \eta \geq 1)$:

$$S(v) \leq \eta S_{unc.}(v) \tag{13.7}$$

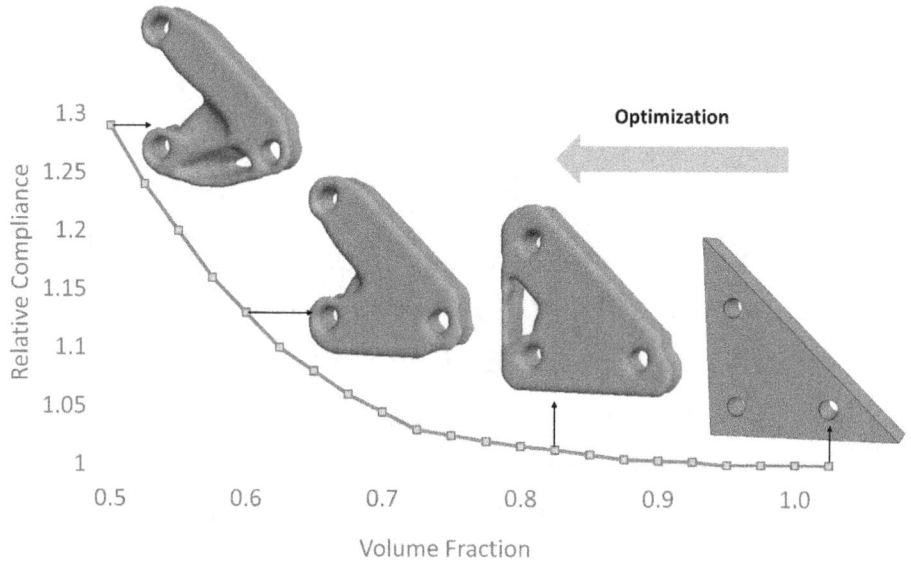

Figure 13.28: Pareto curve for three-hole bracket optimization.

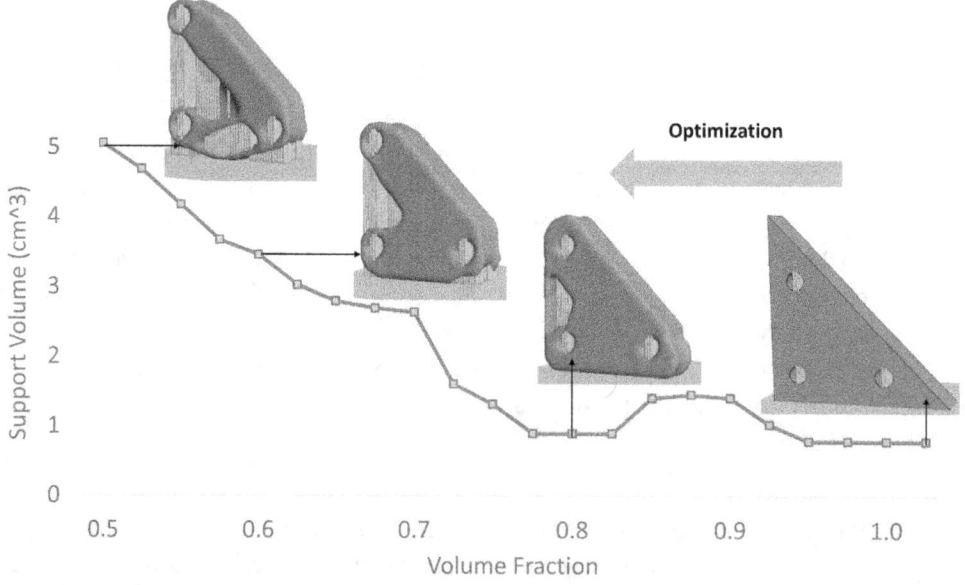

Figure 13.29: Relative support structure volume at different volume fractions for unconstrained problem.

13.5 Topology Optimization for Additive Manufacturing

In other words, Equation 13.7 states that the desired support volume should be less than the unconstrained support volume by a factor of η, at each volume fraction (through interpolation, if necessary). Alternately, one can impose a constraint at the final volume fraction, but imposing a constraint at each volume fraction leads to a smoother optimization process. Further, in this formulation, we treat Equation 13.7 as a *soft* constraint, i.e., the constraint is used to prioritize the solutions within the feasible space (see section 2.5.2), rather than limiting this space. In summary, we propose the following support-structure constrained TO problem, where the parameter η is used to strike a balance between performance and AM costs (see numerical experiments in section 2.8):

$$\begin{aligned}& \underset{\Omega \subset D}{minimize} \{|\Omega|, J\} \\ & S(v) \leq \eta S_{unc.}(v) \quad (soft) \\ & \text{subject to} \\ & \mathbf{Ku} = \mathbf{f}\end{aligned} \quad (13.8)$$

In section 2.5, we consider a gradient based TO framework for solving the above problem. The framework will rely on topological sensitivity for performance, and the proposed topological sensitivity for support structure volume.

Sensitivity analysis

To compute the sensitivity, we begin with the Lagrangian of Equation 13.9:

$$\mathscr{L} = \lambda_V |\Omega| + \lambda_J J + \lambda_S (S(v) - \eta S_{unc.}(v)) \quad (13.9)$$

Thus for topological derivative we have:

$$\mathscr{T}_\mathscr{L} = \lambda_J \mathscr{T}_J + \lambda_S \mathscr{T}_S \quad (13.10)$$

For example, for an intermediate topology in Figure 13.30a, (1) FEA is carried over the new topology, (2) the stresses and strains are computed, and (3) the topological sensitivity field for compliance is known in closed form; the resulting topological sensitivity field is illustrated in Figure 13.30b and Figure 13.30c.

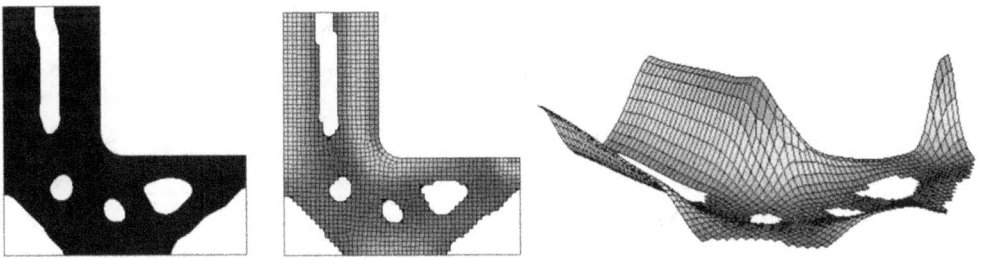

Figure 13.30: (a) Instance of topology, (b) compliance topological sensitivity (2D) (c) 3D view of (b).

Sensitivity of Support Volume based on Surface Angle

Analogous to the topological sensitivity for performance, we propose here topological sensitivity of support structure volume, i.e., *the rate of change in support structure volume with respect to volumetric measure of the hole.* Towards this end, consider the two scenarios illustrated in Figures 13.32 and 13.33, where the design is infinitesimally perturbed either in the interior, or on the boundary.

Interior Hole (Figure 13.32): If a hole of radius ε is inserted in the interior of the domain, one can compute the topological-shape sensitivity as follows. The topological derivative is computed via:

$$\mathcal{T}_S(p \in \Omega) \equiv \lim_{\substack{\varepsilon \to 0 \\ \delta \to 0}} \frac{S(\Omega_{\varepsilon+\delta}) - S(\Omega_\varepsilon)}{V(B_{\varepsilon+\delta}) - V(B_\varepsilon)} \qquad (13.11)$$

In Equation 13.11, $S(\Omega_\varepsilon)$ and $V(B_\varepsilon)$ are support volume and hole volume, for a hole of radius ε. Consider the hole inserted in the interior of the design, we need to find support volume $A = a(A_1 + A_2)$ as illustrated in Figure 13.31.

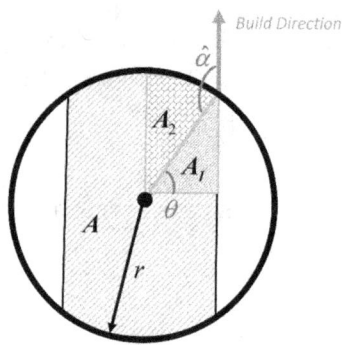

Figure 13.31: Support area in a 2D interior hole.

Since $\theta = \hat{\alpha} - \frac{\pi}{2}$ we have:

$$A_1 = \frac{1}{2}(r\cos(\theta))(r\sin(\theta)) = \tfrac{1}{2}r^2 \sin(\theta)\cos(\theta) = \frac{-1}{2}r^2 \sin(\hat{\alpha})\cos(\hat{\alpha}) \qquad (13.12)$$

$$A_2 = \left(\frac{\frac{\pi}{2} - \theta}{2\pi}\right)\pi r^2 = \frac{(\pi - \hat{\alpha})r^2}{2} \qquad (13.13)$$

$$A = 2r^2(\pi - \hat{\alpha} - \sin(\hat{\alpha})\cos(\hat{\alpha}) \qquad (13.14)$$

Next to find the support volume in a spherical ball with radius r we extend Equation 13.14 as follows:

$$\begin{aligned} S &= \int_{-r\cos(\theta)}^{r\cos(\theta)} 2(r^2 - x^2)(\pi - \hat{\alpha} - \sin(\hat{\alpha})\cos(\hat{\alpha}))dx \\ &= 4r^3(\pi - \hat{\alpha} - \sin(\hat{\alpha})\cos(\hat{\alpha})\left(\sin(\hat{\alpha}) - \frac{\sin^3(\hat{\alpha})}{3}\right) \end{aligned} \qquad (13.15)$$

13.5 Topology Optimization for Additive Manufacturing

Finally based on Equation 13.11 the topological sensitivity is computed via Equation 13.16:

$$\mathcal{T}_S(p \in \Omega) = \lim_{\substack{\varepsilon \to 0 \\ \delta \to 0}} \frac{4r^3(\pi - \hat{\alpha} - \sin(\hat{\alpha})\cos(\hat{\alpha}))(\sin(\hat{\alpha}) - \frac{\sin^3(\hat{\alpha})}{3})((\varepsilon+\delta)^3 - \varepsilon^3)}{\frac{4}{3}\pi((\varepsilon+\delta)^3 - \varepsilon^3)} \quad (13.16)$$

i.e.

$$\mathcal{T}_S(p \in \Omega) = \frac{3(\pi - \hat{\alpha} - \sin(\hat{\alpha})\cos(\hat{\alpha}))(\sin(\hat{\alpha}) - \frac{\sin^3(\hat{\alpha})}{3})}{\pi} \quad (13.17)$$

Where $\frac{\pi}{2} \leq \hat{\alpha} \leq \pi$ is the threshold angle. For example, if the threshold angle $\hat{\alpha} = \frac{\pi}{2}$, then $\mathcal{T}_S(p) = 1$, i.e., the entire hole will need to be filled with support structures; a typical value is $\mathcal{T}_S(p \in \Omega) \approx 0.72$ when $\hat{\alpha} = \frac{3\pi}{4}$.

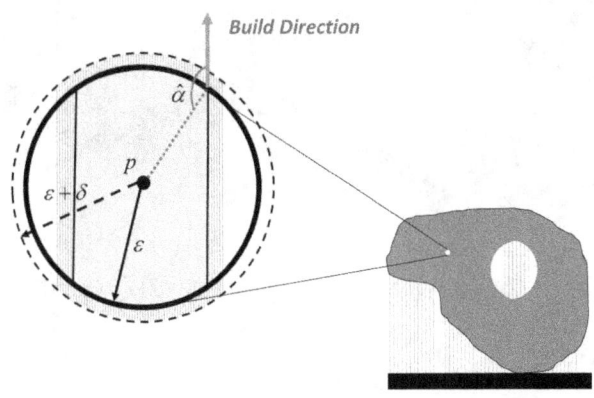

Figure 13.32: Sensitivity of support volume in the interior.

Boundary Hole (Figure 13.33): Unlike the interior, the support volume on the boundary depends both on the local neighborhood (curvature) and the length and direction of support. In order to capture both, we define a scalar function $F^S(x_p)$ at each boundary point as follows:

$$F^S(x_p) = \frac{1}{2}l_p(1 - \cos(\alpha_p)) \quad (13.18)$$

In Equation 13.18, α_p is the angle between surface normal and build direction at boundary point p. We compute the sensitivity for the worst-case scenario, where boundary is perturbed along support at each point $\hat{S}_p$. One can then show that the sensitivity at the boundary is given by Equation 13.19:

$$\mathcal{T}_S(p \in \partial\Omega) = \frac{1}{2}(1 - \cos(\alpha_p)) \quad (13.19)$$

Further, for each overhang point, the same sensitivity value is assigned to its corresponding opposite point (see Figure 13.33).

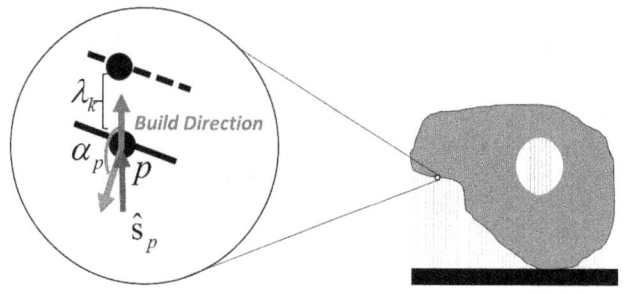

Figure 13.33: Sensitivity of support volume on boundary.

Given the above definitions, one can compute the support volume sensitivity at all points; this is illustrated in Figure 13.34a and Figure 13.34b.

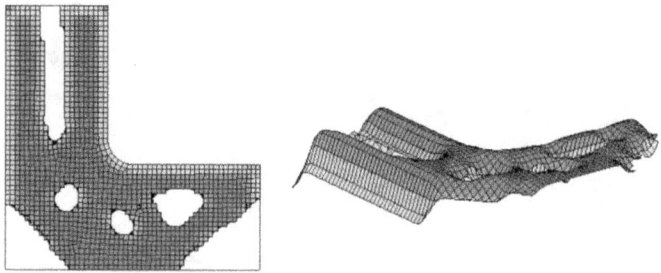

Figure 13.34: (a) Sensitivity of support volume (2D). (b) 3D view of (a).

Sensitivity Weighting

Once the performance and support volume sensitivities are computed and normalized, we exploit the well-established augmented Lagrangian method [13.1] to impose support structure constraint. Specifically, the support-constrained in Equation 13.8 is first expressed in the standard form:

$$g = \begin{cases} \dfrac{S(v)}{\eta S_{unc.}(v)} - 1 \leq 0 & \eta S_{unc.}(v) \neq 0 \\ S(v) = 0 & \eta S_{unc.}(v) = 0 \end{cases} \tag{13.20}$$

A popular method for imposing such constraints the augmented Lagrangian method [13.1], where the constraint and objective are combined to a single field:

$$\mathscr{L} = J + \mathscr{L}_g \tag{13.21}$$

where:

$$\mathscr{L}_g = \begin{cases} \lambda g + \dfrac{1}{2}\gamma(g)^2 & \lambda + \gamma g > 0 \\ \dfrac{1}{2}\dfrac{\lambda^2}{\gamma} & \lambda + \gamma g \leq 0 \end{cases} \tag{13.22}$$

13.5 Topology Optimization for Additive Manufacturing

Where λ is the Lagrangian multiplier and γ is the penalty parameter (that are updated during the optimization process [13.1]). By taking the topological derivative of Equation 13.21, we arrive at Equation 13.23 for the effective sensitivity [13.2]:

$$\mathcal{T} = \mathcal{T}_J + w_S \mathcal{T}_S \qquad (13.23)$$

where

$$w_s = \begin{cases} \mu + \gamma g & \mu + \gamma g > 0 \\ 0 & \mu + \gamma g \leq 0 \end{cases} \qquad (13.24)$$

Observe that the weight on the support structure sensitivity is zero if $g < \dfrac{-\lambda}{\gamma}$, else it takes a positive value.

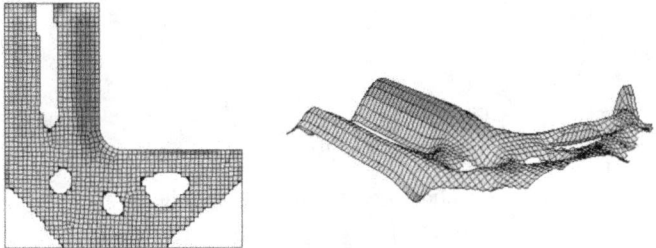

Figure 13.35: (a) Equally weighted sum of the two sensitivity fields (2D) (b) 3D view of the sensitivity field.

To illustrate Equation 13.23, suppose the two topological sensitivity fields are normalized to unity, and suppose $w_s = 1.0$, the resulting field is illustrated in Figure 13.35a and Figure 13.35b. Observe that the resulting field is a combination of the two fields in Figure 13.30 and Figure 13.34. As the optimization progresses, the weight is determined dynamically from Equation 13.24, while the parameters μ and γ are updated during each iteration as described in [13.2].

Limitations
Since the topological sensitivity depends on the *rate of change* in support structure volume, to ensure convergence of the optimization process it must necessarily be smooth and differentiable. However this is not the case in practice and in some cases a slight change in surface angle might introduce or eliminate a large volume of support structure. Topological sensitivity on boundary points proposed in Equation 13.19 is a sufficiently smooth approximation to what is expected in practice. Hence this assumption on differentiability prevents the current algorithm to make decisions that require abrupt change in support volume. This becomes highlighted in special cases where it is possible to remove all support structure through introducing certain shapes such as isosceles. This will be demonstrated in Numerical experiments.

Algorithm
Piecing these concepts together, the algorithm proceeds as follows (see Figure 13.36):

Algorithm 13.5.1 **Support constrained Pareto tracing**

1. It is assumed that the unconstrained optimization problem has been solved, and $S_{unc.}(v)$ has been computed.
2. Carry out FEA on Ω; compute the normalized sensitivity fields $\mathcal{T}_J$, $\mathcal{T}_S$ and the weighted field $\mathcal{T}$ as described above; smoothen the $\mathcal{T}_J$ field. Observe that, every time the topology changes, FEA must be executed and the topological sensitivities recomputed.
3. Treating $\mathcal{T}_J$ as a level-set function, extract a new topology Ω using fixed-point iteration, and the iso-surface is extracted. If the topology has not converged, repeat steps 2 and 3.
4. Decrement the volume fraction and return to step 2 until the desired volume is reached.

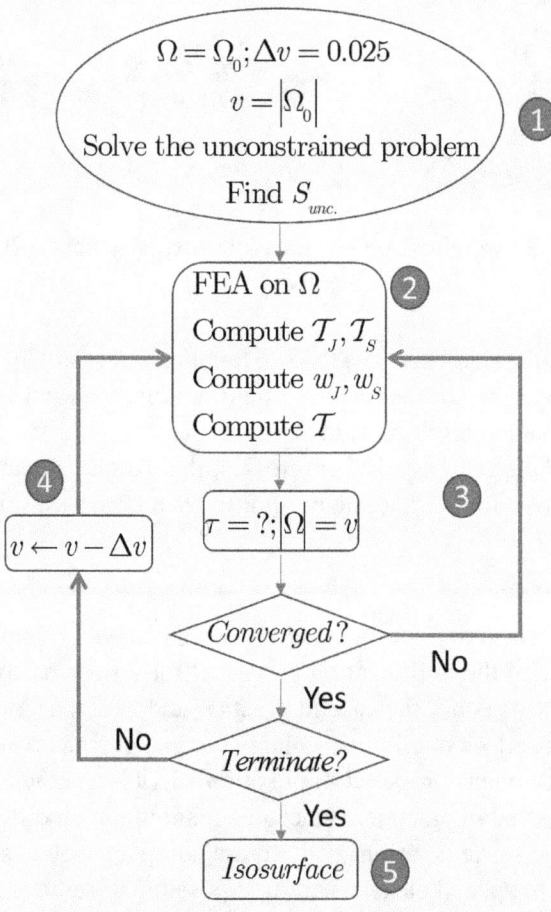

Figure 13.36: Support Structure constrained Pareto algorithm.

13.5 Topology Optimization for Additive Manufacturing

Numerical Experiments

In this section, we demonstrate the proposed method through several examples. In all of the experiments, the material is assumed to be isotropic ABS plastic with Young's modulus of $E = 2GPa$ and Poisson ratio of $v = 0.39$. The threshold angle $\hat{\alpha}$ is assumed to be $\frac{3\pi}{4}$, unless otherwise noted.

2D MBB

Consider the 2D MBB design (implicit thickness of 1 cm) in Figure 13.37 whose support structure reduction was studied in Figure 13.24b. The initial design requires no support and the objective is to find stiffest design at 0.65 volume fraction.

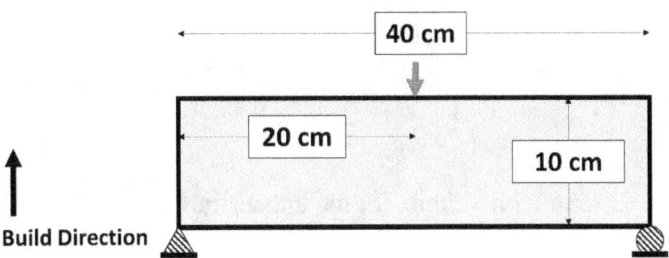

Figure 13.37: 2D MBB example with boundary conditions and build direction.

Recall that we first solve the unconstrained problem, and a series of topologies that lie on the Pareto curve are generated; see Figure 13.38. Figure 13.39 illustrates the corresponding support volume in cm3.

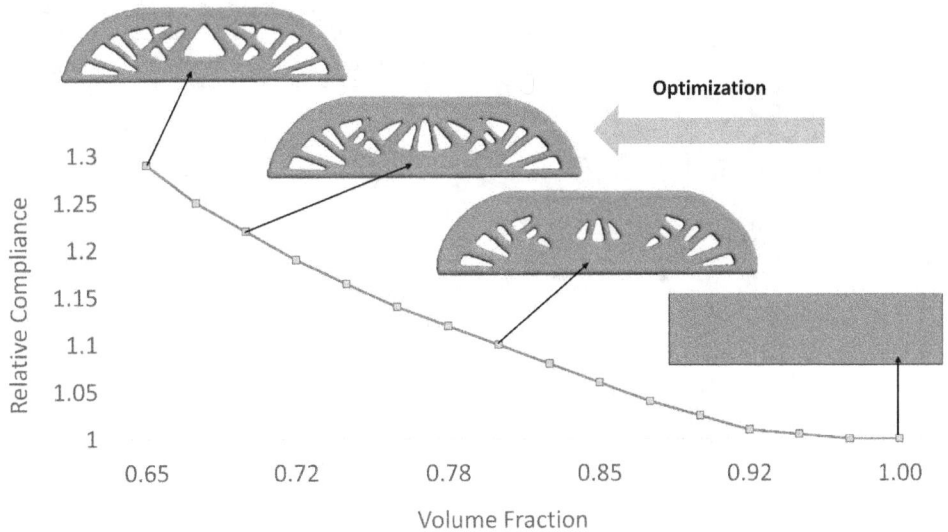

Figure 13.38: Compliance Pareto curve for the MBB beam.

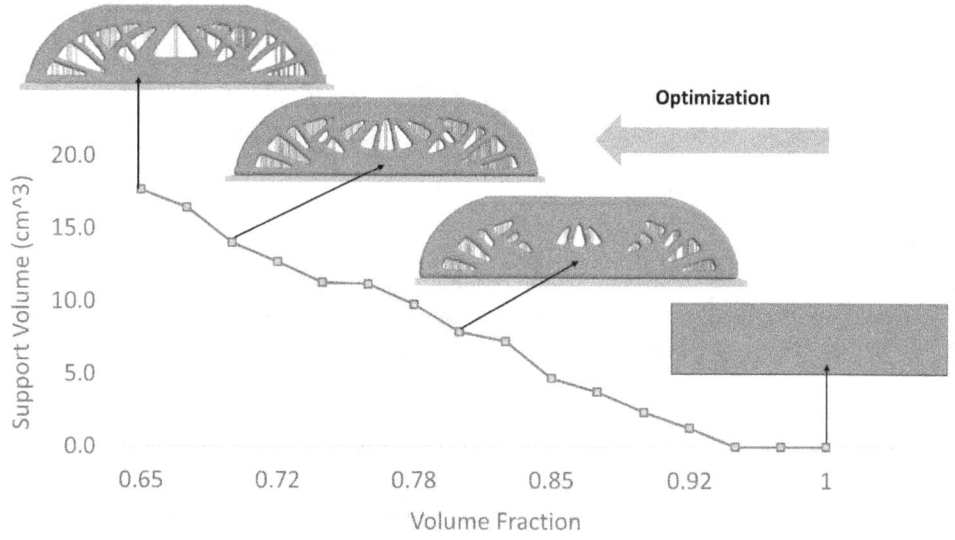

Figure 13.39: Support volume for the unconstrained MBB beam problem.

The unconstrained support volume from Figure 13.39 is then used as a reference to impose a support structure constraint. In particular, we study the impact of the relative constraint η (see Equation 13.8) on the *final topology* at a volume fraction of 0.65. Table 13.3 summarizes the results; observe that with increased support structure constraint, the proposed method reduces the number of internal holes. This is, by no means, the unique solution to the problem; it happens to be a solution that meets the desired constraints.

Table 13.3: 2D MBB. Effect of support constraint on optimized design.

Final Topology	Support Volume Constraint	Support Volume Achieved	Relative Compliance
	-	-	1.29
	80%	62%	1.34
	60%	59%	1.42
	40%	42%	1.56
	0%	0%	1.75

13.5 Topology Optimization for Additive Manufacturing

In this particular example changing shape of the holes to isosceles would be more desirable, but that would require abrupt change in sensitivity analysis and current framework avoids such designs.

Three-Hole Bracket

In this example, we study the impact of the support structure constraint over the entire Pareto curve. Considering the three-hole bracket illustrated earlier in Figure 13.27 and recalling the compliance Pareto curve for the unconstrained problem in Figure 13.28, and the corresponding support structure curve in Figure 13.29; Figure 13.40a illustrates compliance Pareto curves for the unconstrained and the constrained case. As expected, imposing the support constraint increases compliance.

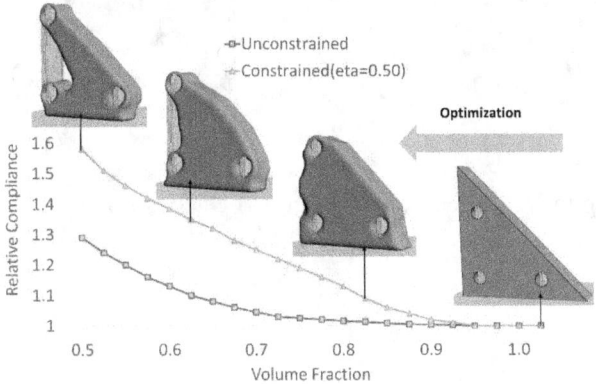

(a) Evolution of compliance

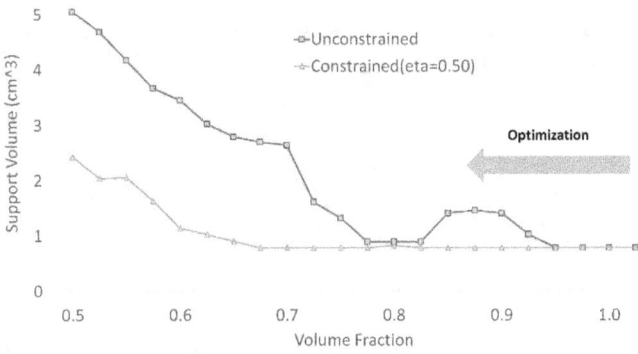

(b) Evolution of support volume

Figure 13.40: Unconstrained and constrained Pareto curves for three-hole bracket optimization.

Figure 13.40b illustrates the evolution of support structure volume for the two scenarios. Observe that as expected, removing more material can either increase or decrease the support volume due to its nonlinearity, nonetheless imposing a stringent constraint on support structure consistently reduces the support volume w.r.t the corresponding unconstrained design.

The support volume prior to optimization is $S_0 = 0.79(cm^3)$. The objective is to find stiffest design at 0.5 volume fraction. Figure 13.41 illustrates the optimized design for (a) unconstrained, (b) constrained with $\eta = 0.5$. Relative compliance values for these cases are respectively 1.24 and 1.58.

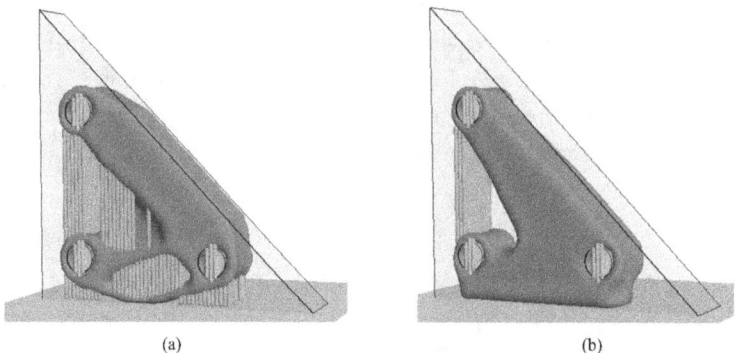

Figure 13.41: Optimized three-hole bracket. (a) Unconstrained (b) Constrained with $\eta = 0.50$.

Mount Bracket

Consider the mount bracket of Figure 13.42 subject to structural constraints and loading as illustrated. The threshold angle $\hat{\alpha}$ is assumed to be $3\pi/4$. The build direction illustrated in Figure 13.42 is chosen such that it gives the best surface quality on the larger cylindrical face; for this design, prior to optimization the support volume is $S_0 = 1.12(cm^3)$. The objective is to find stiffest design at 0.7 volume fraction.

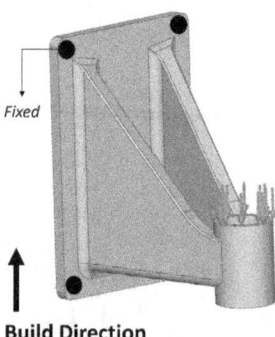

Figure 13.42: Mount bracket with boundary conditions and build direction.

Figure 13.43 illustrates the optimized designs of (a) unconstrained and (b) constrained with $\eta = 0.80$. The final support structure volume for the unconstrained design is $9.24(cm^3)$ while for the constrained design it has reduced by about 17% to $7.70(cm^3)$.

13.5 Topology Optimization for Additive Manufacturing

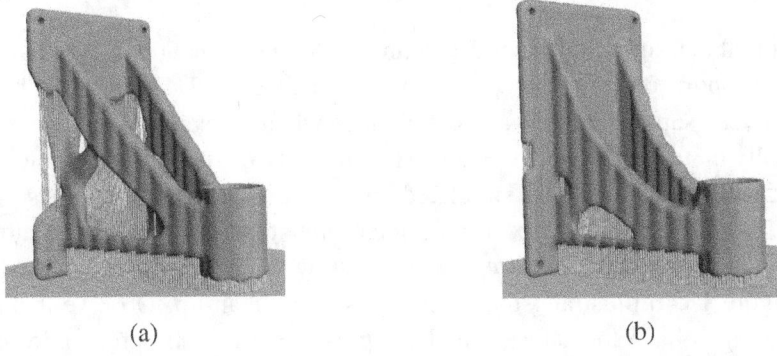

(a) (b)

Figure 13.43: Optimized mount bracket at 0.7 volume fraction (a) Unconstrained (b) constrained with 0.8 support fraction.

Figure 13.44 illustrates the evolution of support volume throughout the optimization process. Observe that up to 0.9 volume fraction the unconstrained and constrained results are very similar. However for lower volume fractions the constrained support volume is consistently about 20% smaller than that of unconstrained design.

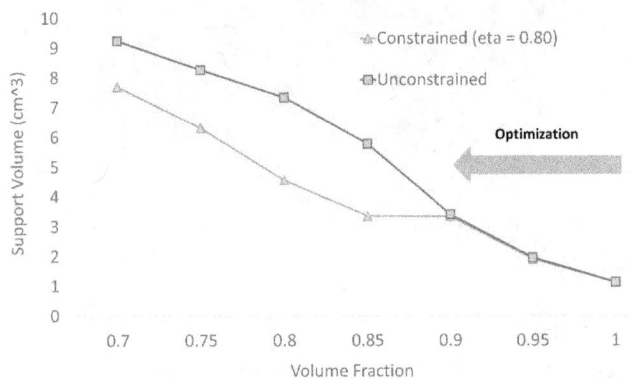

Figure 13.44: Evolution of support volume for the mount bracket.

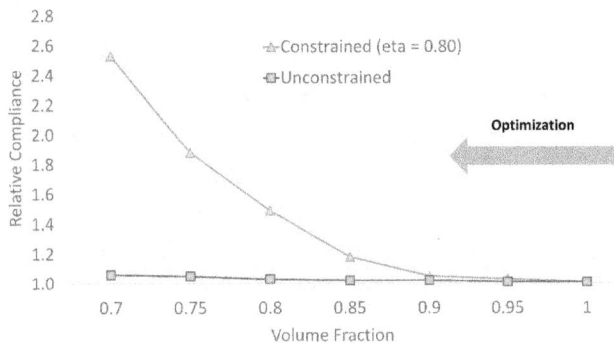

Figure 13.45: Evolution of compliance for the mount bracket.

Figure 13.45 illustrates the evolution of relative compliance values as more material is removed from the design. For the unconstrained design the final C/C_0 is about 1.05, while by imposing support constraint this value increases to about 2.52. Figure 13.45 highlights the trade-off between support volume and compliance when the support constraint is imposed. It is essentially up to the designer to choose the intensity of support constraint.

To verify the validity of these simulated results, each of these topologies were printed on an *XYZ Da Vinci 2.0* fused deposition printer. Note that the support structures were not generated by our algorithm, they were introduced by the XYZ software, based on default settings. Figure 13.46 illustrates the actual parts after clean-up. Observe that both of the optimized designs have the same weight (as prescribed by the optimization), while the amount of support structure is substantially reduced in the constrained design. This example illustrates the effectiveness of the proposed algorithm in handling support constraints.

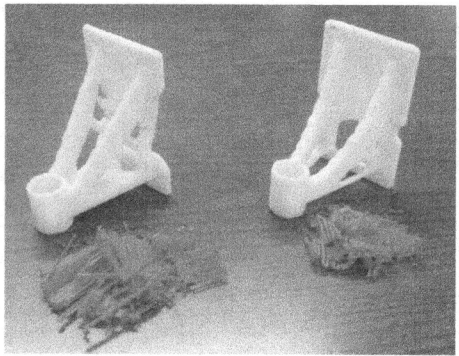

Figure 13.46: Printed mount bracket and the required support structures at 0.7 volume fraction.

Different Build Directions

In this section, we demonstrate the robustness of the proposed method with respect to the build directions. Consider the problem posed in Figure 13.47 where the geometry is described via numerous curved surfaces and two cylindrical holes in two different directions; this makes picking the optimal build orientation challenging. Further to capture the complexity of the design, a hexahedral mesh with about 1.7 million degrees of freedom was used.

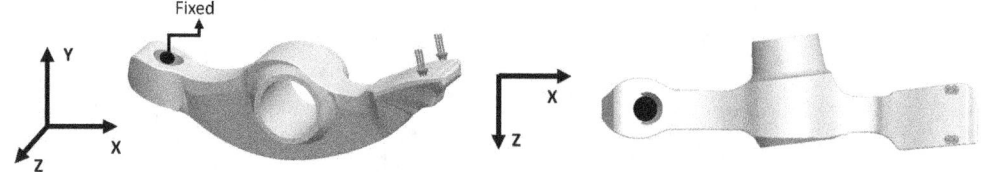

Figure 13.47: Rocker arm of Honda Supra-X 100 cc (grabcad.com): (a) Iso view, (b) Top view.

A plausible choice for the build direction is -Z, as shown in Figure 13.48. In this direction, the larger cylinder has better surface quality and the initial support is minimal. First, we optimize the design for minimum compliance at 0.7 volume fraction without imposing any constraints on support structure(Figure 13.48a). In this particular orientation, $S_{unc.}(0.7)$

is smaller than S_0, which means that during optimization, some of the overhanging surfaces are removed to reduce the overall support volume. Next, in order to further reduce support structure, we set $\eta = 0.90$ and solved the optimization problem of Equation 13.8 to arrive at the design in Figure 13.48b. Observe that by imposing the support constraint, no additional overhangs are created, however since the initial design is dominant, support volume is reduced by only about 3%, while the compliance has increased by about 15%.

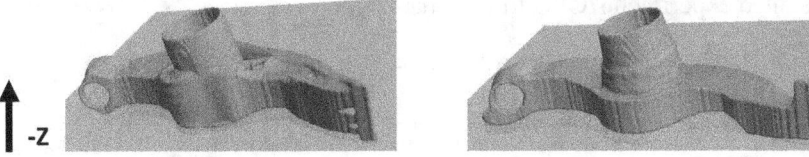

Figure 13.48: Rocker arm. Building in -Z direction a) unconstrained b) constrained.

Next, the build direction was set to +Y since it gives better surface quality for the smaller cylindrical hole. Solving the same optimization problem as before results in the unconstrained design in Figure 13.49a and constrained design in Figure 13.49b with $\eta = 0.90$. The support volume was reduced by 20%, while the compliance increased by 32%.

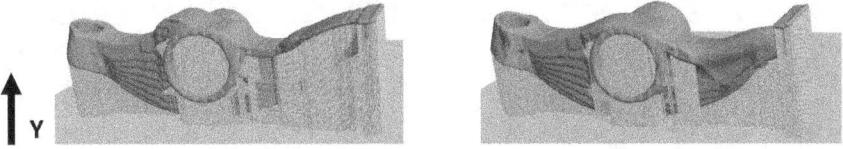

Figure 13.49: Build direction along +Y direction: (a) unconstrained, and (b) constrained.

Finally, the build direction was set to +X; a justification for this direction can be better fusion between layers, since the print area is smaller than previous directions. The results are summarized in Figure 13.50: the support volume was reduced by 4%, while the compliance increased by 10%.

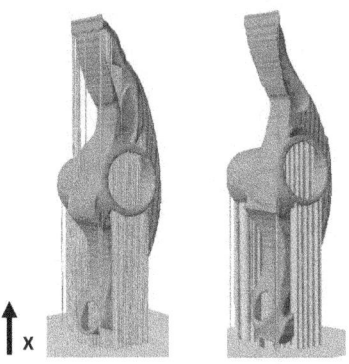

Figure 13.50: Rocker arm. Building in +X direction, unconstrained(left) and constrained (right)

Computational Cost

In this section, we study the convergence and performance of the proposed algorithm. All experiments are conducted on a Windows 7 64 bit machine with an 8-core Intel Core i7 CPU running at 3.00 GHz, and 16 GB of memory.

Table 13.4 summarizes the CPU times of the unconstrained and constrained examples. Observe that as the size of the problem and the support volume increases, the constrained problem requires more computational effort to compute support sensitivity field, yet for all of the presented experiments CPU time remains comparable.

Table 13.4: Computational cost, with and without support structure constraints.

Example	Finite element degrees of freedom	CPU time Unconstrained	CPU time Support Constrained
MBB	27,400	5.25 sec.	5.5 sec.
Three-hole bracket	45,000	10 sec.	$(\eta = 0.75)$ 11 sec. $(\eta = 0.50)$ 13.7 sec.
Rocker Arm (-Z)	$\approx$ 1.7 million	28 min 30 sec.	30 min 59 sec.
Rocker Arm (+Y)	$\approx$ 1.7 million	28 min 30 sec.	32 min 6 sec.
Rocker Arm (+X)	$\approx$ 1.7 million	28 min 30 sec.	30 min 14 sec.

13.6 References

Reference 13.1 J. Nocedal and S. Wright, *Numerical Optimization*, Springer Science & Business Media, 2006.

Reference 13.2 S. Deng and K. Suresh, *Multi-constrained topology optimization via the topological sensitivity*, Structural and Multidisciplinary Optimization, vol. 51, no. 5, pp. 987–1001, Nov. 2014.

Reference 13.3 D. Brackett, I. Ashcroft, and R. Hague, *Topology optimization for additive manufacturing*, in 22nd Annual international solid freeform fabrication symposium, 2011, pp. 348–362.

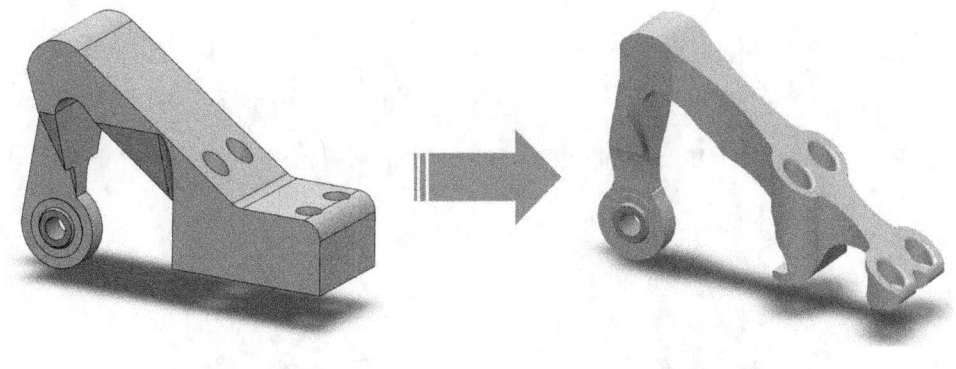

14. Case Studies

In this chapter, we will discuss couple of industrial topology optimization case studies.

14.1 GE-GrabCAD

The first case-study is the GE-GrabCAD multi-load problem in Figure 14.1. This is a multi-load problem subject to stress constraint.

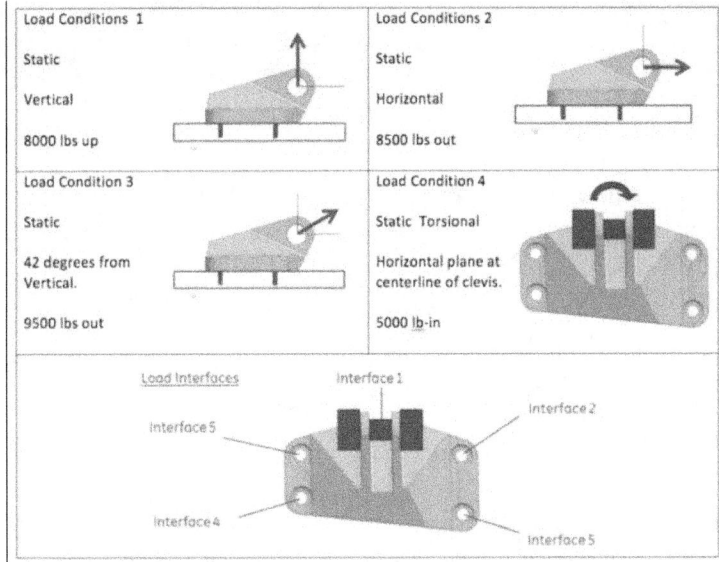

Figure 14.1: GE-GrabCAD design problem
(https://grabcad.com/challenges/ge-jet-engine-bracket-challenge)

Example 14.1 GE-GrabCAD design problem

1. Load the GEGrabCADProject.prj from the ParetoWinExamples folder.

   (R) Observe that the yield stress (stress limit) is 120 ksi.

2. Use the Load Set option under *Boundary Conditions* menu to scroll through all the four loads.
3. The four different loads are illustrated in Figure 14.2.

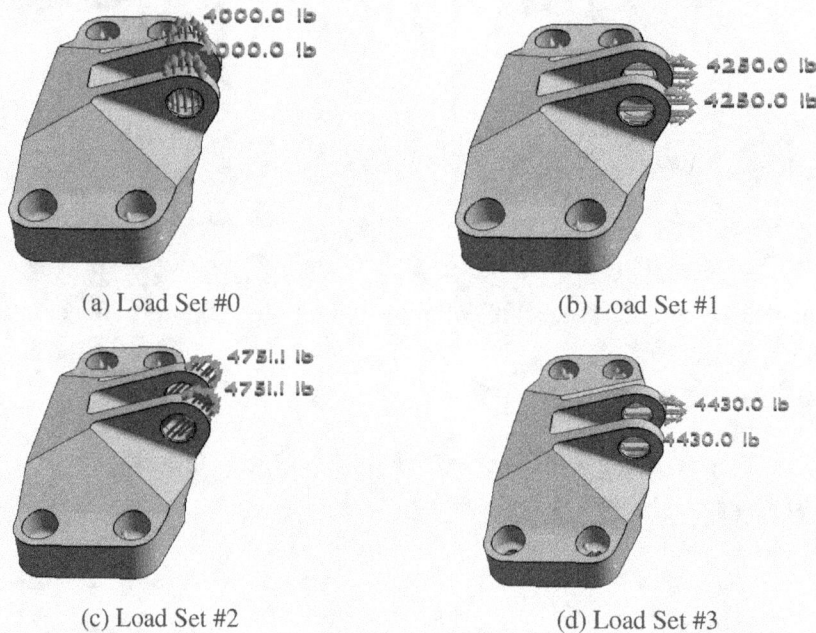

Figure 14.2: The four loads.

4. Next, using a fine quality mesh, carry out FEA for each of the loads; see Figure 14.3. The maximum stresses for the four loads are 31 ksi, 34 ksi, 24 ksi, and 36 ksi respectively.

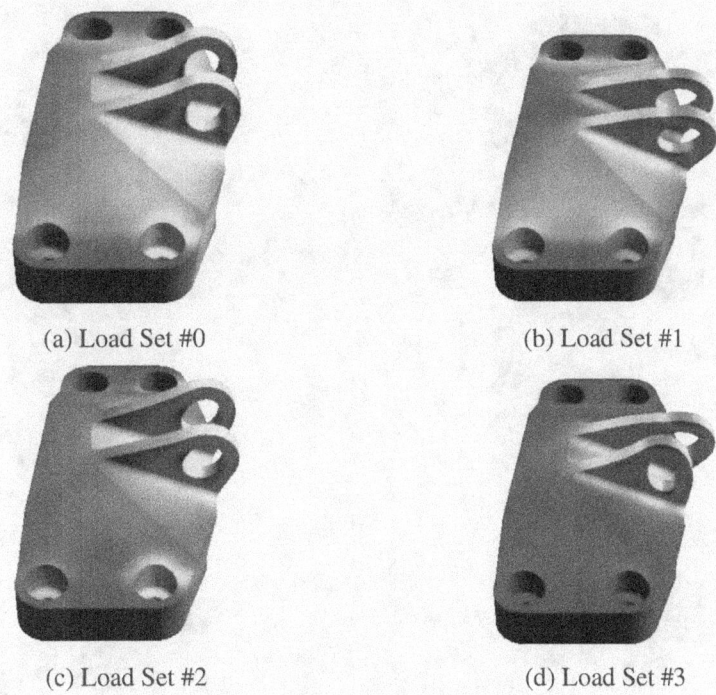

(a) Load Set #0 (b) Load Set #1

(c) Load Set #2 (d) Load Set #3

Figure 14.3: Finite element analysis.

5. The optimization parameters are according to the project file; we will remove as much material as possible, while satisfying the (yield) stress constraint (the displacement constraint can be disregarded for this project).
6. The optimized model of 0.21 volume fraction is illustrated in Figure 14.4.

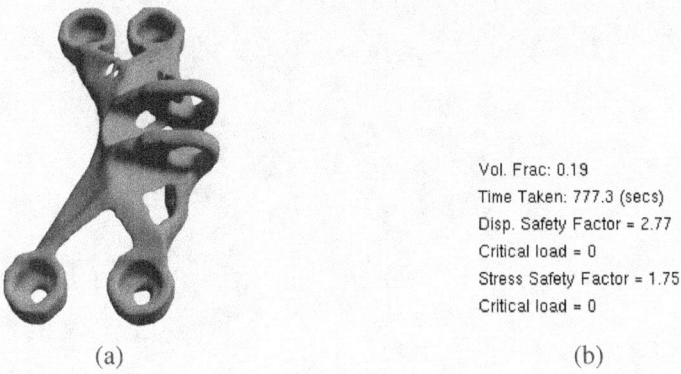

(a) (b)

Figure 14.4: Optimized model 0.21 volume fraction.

7. Next, carry out FEA on the optimized model, for each of the loads, by first unselecting the *Remesh* option.
8. The final stresses are 124 ksi, 41.4 ksi, 101 ksi and 93.2 ksi respectively. Observe that Load Set #0 is the dominating load.

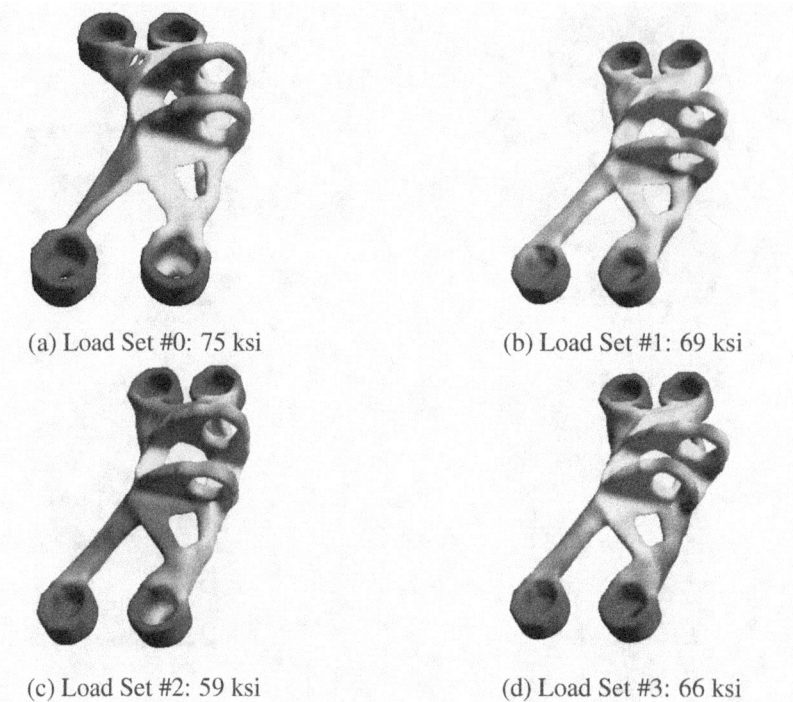

(a) Load Set #0: 75 ksi

(b) Load Set #1: 69 ksi

(c) Load Set #2: 59 ksi

(d) Load Set #3: 66 ksi

Figure 14.5: FEA on the optimized model, and maximum stress under each load.

14.2 Alcoa-GrabCAD

The second case-study is the Alcoa-GrabCAD multi-load problem in Figure 14.6.

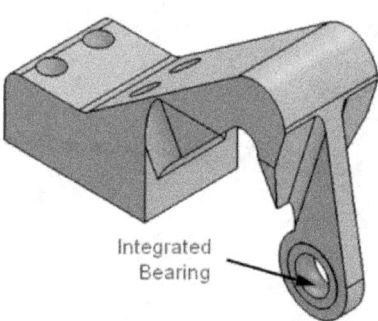

Figure 14.6: Alcoa-GrabCAD design problem.

■ **Example 14.2 Alcoa-GrabCAD design problem**

1. Load the *AlcoaGrabCADProject.prj* from the *ParetoWinExamples* folder.
2. Set units to IPS.

3. The material associated with the project. Observe that the yield stress (stress limit) is 150 ksi.
4. Use the *Load Number* option under *Structural Loads* menu to scroll through all the three loads.
5. The three different loads are illustrated in Figure 14.7.

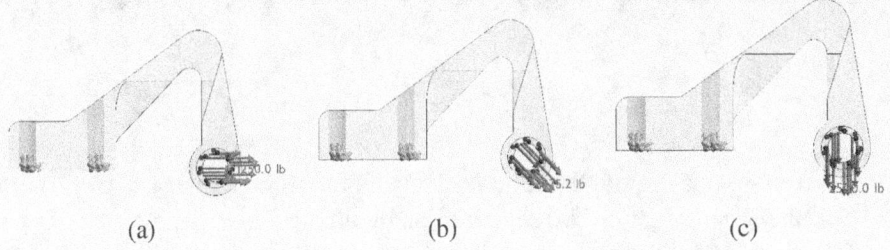

(a) (b) (c)

Figure 14.7: The three loads.

6. Next, using a fine quality mesh, carry out FEA for each of the loads; see Figure 14.8. The maximum stresses for the three loads are 106 ksi, 101 ksi, and 91 ksi respectively.

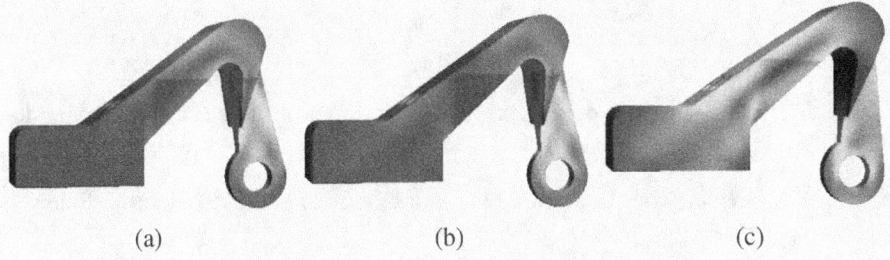

(a) (b) (c)

Figure 14.8: Finite element analysis for the three loads.

7. The optimization parameters are as follows:
   - Set desired volume fraction to 0.10; we will remove as much material as possible, while satisfying the (yield) stress constraint.
   - Allowable displacement is 0.10 in.
   - Impose draw-direction along Z axis.
   - Toggle on *Keep Fixed Faces*.
   - Impose symmetry along Z axis (under *FEA* menu).
8. The optimized model of 0.30 volume fraction is illustrated in Figure 14.9.

(a)   (b)

Figure 14.9: Optimized Alcoa model.

9. Next, carry out FEA on the optimized model, for each of the loads, by first _(toggle off) the *Remesh* option and setting mesh quality to *Fine*.
10. The final stresses are respectively: 127 ksi, 121 ksi and 139 ksi.

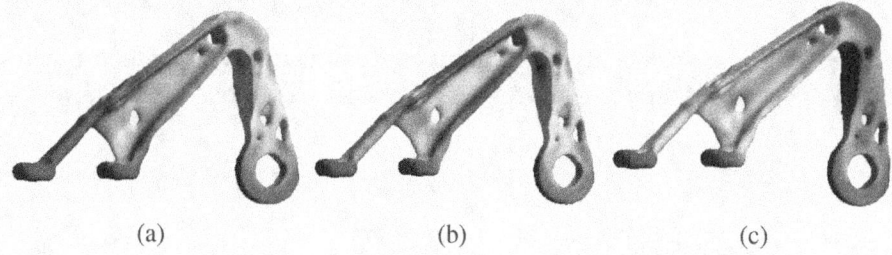

(a)   (b)   (c)

Figure 14.10: FEA on the optimized model, and maximum stress under each load.

   Ⓡ   Observe that Load Set #2 (Figure 14.10c) is the dominating load.

15. Appendix: Pareto Menus

For quick reference, we will summarize here the menu options within Pareto.

15.1 Units

Under *Units* in Figure 15.1, we have the following options:

- **Dimensions and forces**: There are three choices available for dimensional and structural loading units:
    1. $[\text{MKS}]$ (m, N, s)
    2. $[\text{mmKS}]$ (mm, N, s)
    3. $[\text{IPS}]$ (in, lb, s)
- **Temperature**: There are three choices available for temperature units:
    1. Celsius (°C)
    2. Kelvin (K)
    3. Fahrenheit (°F)
- **Angle**: There are two options available for angular units:
    1. Degree
    2. Radian

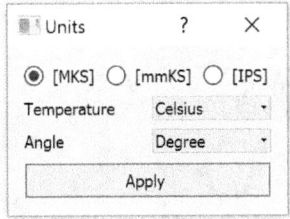

Figure 15.1: Units menu.

15.2 Geometry

Under *Geometry* menu in Figure 15.2 we have:

- **Load stl**: to replace the current model with a new geometry and start afresh.
- **Update stl**: to replace the current model with a new geometry while copying over the current boundary conditions. This assumes that the current model and new model are similar, and that the boundary conditions can be mapped without ambiguity.

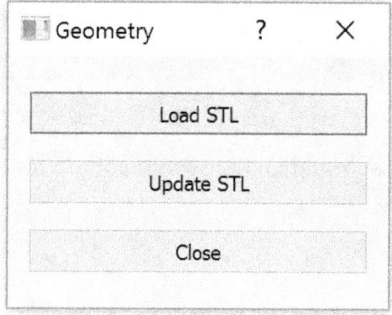

Figure 15.2: Geometry menu.

15.3 Material

In *Material* menu, in 15.3a we have the following choices:

- **Material**: select one of the pre-defined materials or create a custom material.
- **E**: modify Young's modulus.
- **nu**: modify Poisson ratio.
- **Y**: modify yield strength.
- **Density**: modify density.
- **K**: modify thermal conductivity.
- **Cp**: modify specific heat.
- **Alpha**: modify thermal expansion coefficient.
- **Do not Optimize?**: specify whether this model is being optimized.

Figure 15.3b shows the list of pre-defined materials.

15.4 Loads

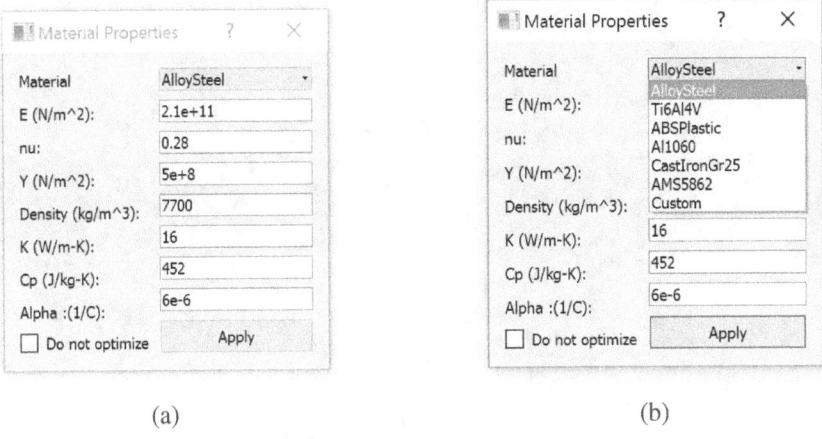

Figure 15.3: (a) Material menu, (b) List of materials.

15.4 Loads

Under the *Loads* menu in Figure 15.4, we can specify:
- **Load Number**: this is useful when there are more than one loading conditions defined.
- **Selection**: helps in selecting planes, cylinders and individual triangles.
- **Load Type**: determines the boundary condition to be imposed on the selected triangles.

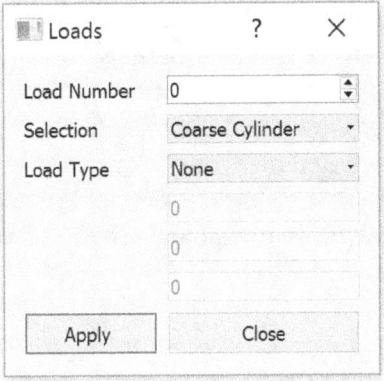

Figure 15.4: Loads menu.

The *Selection* option specifies which STL triangles must be chosen when a part of the model is clicked on:
- **Triangle**: selects a single triangle.
- **Plane**: selects all of the triangles that lie on the same plane as the triangle selected.
- **Fine Cylinder**: selecting all triangles that lie on a cylindrical surface such that normal of adjacent triangles are less than 10 degrees.
- **Coarse Cylinder**: selecting all triangles that lie on a cylindrical surface such that normal of adjacent triangles are less than 15 degrees.
- **Other**: selecting all triangles that lie on a cylindrical surface such that normal of adjacent triangles are less than 20 degrees.

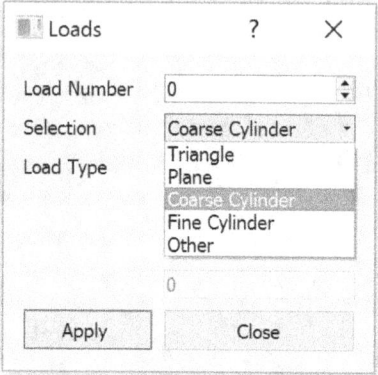

Figure 15.5: Face select options.

Once a region of the geometry is selected, there are different options for imposing loads as illustrated in Figure 15.6.

- **None**: remove the current BC on the selected faces.
- **XYZFixed**: restrain selected faces in all directions.
- **XFixed**, **YFixed**, **ZFixed**: restrain selected faces only in x, y, or z directions, respectively.
- **Force**: apply a general force on the selected faces by specifying x, y, and z components.
- **NormalForce**: apply a force normal to selected faces.
- **Pressure**: apply a pressure on selected faces.
- **Torque**: apply a torsional loading on selected faces.
- **Sliding**: restrain the selected faces only along their normal, while allowing planar movement.
- **Retain**: this option is specific to topology optimization, where materials from this region must not be removed.

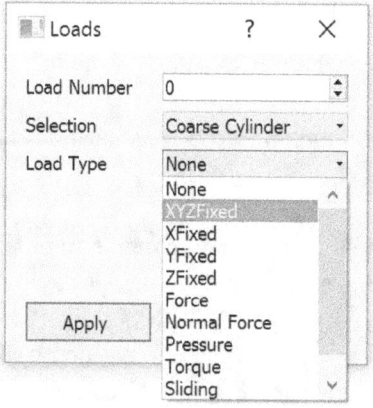

Figure 15.6: Load types.

15.5 Body Force

It is also possible to impose uniform body force on the model, for instance to capture gravitational effects, by specifying the three components in x, y, and z directions, as in Figure 15.7.

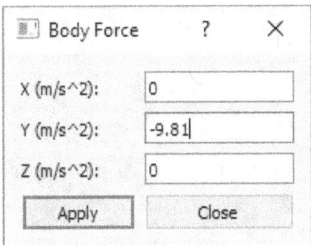

Figure 15.7: Acceleration loads (body forces).

15.6 Display

The *Display* menu in Figure 15.8 offers many options for examining the geometries and solution fields.

- **Geometry**: specifies the geometry to use (initial design or final topology)
- **Field**: specifies the field to display (none, displacement, stress, ...).
- **Show bounding box**: display the bounding box?
- **Show triangles**: display the individual triangles within the STL model?
- **Scale deformation**: should the deformation be scaled for convenient visualization?
- **Show transparent geometry**: display the initial model as transparent?
- **Show axis**: display axes of the coordinate system?
- **Show structural loads**: display the elasticity loadings applied on the model?
- **Show symmetry planes**: display the planes about which the model is symmetric?
- **Show non-design parts**: display the parts that are not being optimized?
- **Animate for 3 cycles**: animate the static solution field for 3 cycles.

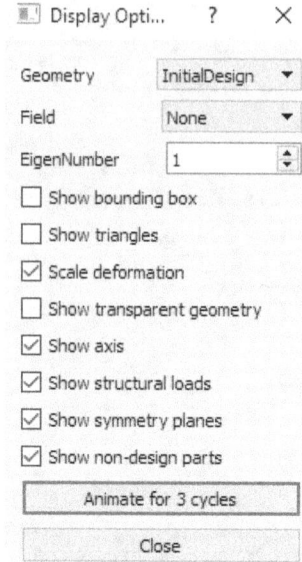

Figure 15.8: Display menu.

15.7 Finite Element Analysis

Once a valid structural problem is defined, we can use *FEA* menu to carry out a finite element analysis and study displacement and stress fields. Figure 15.9 shows different options where:

- **Load**: current active load set for this FEA.
- **Mesh Quality**: quality of the underlying finite element mesh:
    1. **Very Coarse**: about 10,000 elements.
    2. **Coarse**: about 25,000 elements.
    3. **Medium**: about 50,000 elements.
    4. **Fine**: about 100,000 elements.
    5. **Very Fine**: about 250,000 elements.
- **#Elements**: Specify the number of elements.
- **#Modes**: Number of modes to compute (see below).
- **Remesh**: create a new finite element mesh before solving the FEA.
- **X symmetry, Y symmetry**, and **Z symmetry**: mesh must be symmetric along x, y, and z, respectively.
- **Static Structural**: Solve the static FEA problem.
- **Modal Structural**: Solve the modal problem.

15.8 TopOpt Constraints

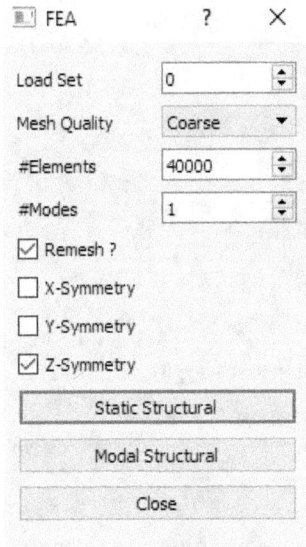

Figure 15.9: Finite Element Analysis menu.

15.8 TopOpt Constraints

Through the constraint menu in Figure 15.10, we can impose various types of performance and manufacturing constraints on the optimization process.

- **Draw Direction**: impose casting constraint in the specified direction, so that there would be no cavities in that direction.
- **Through Cut**: impose laser cut constraint so that the cross-section remains constant along the specified direction.
- **Cyclic Sym (Z)**: impose cyclic symmetry in X-Y plane
- **RelMinFeatSize**: specify minimum feature size that can be manufactured.
- **Stress Safety Factor**: allowable stress safety factor, based on material yield strength.
- **Upperlimit Disp.**: allowable displacement.
- **Lowerlimit 1st eigenmode.**: lowerlimit on the first eigen-mode (if negative, then this constraint is not imposed).
- **Keep Fixed Faces**: whether all the restrained faces must be retained.
- **Apply**: Apply the constraints.

Figure 15.10: Topology optimization constraints.

15.9 Topology Optimization

Under *Optimize* menu in Figure 15.11:

- **Objective**: The objective to minimize compliance, minimize stress or maximize the first eigen-mode.
- **Use All Loads?**: optimize the model considering all of the imposed loading conditions.
- **Use Load**: specify which individual load case must be considered for this optimization.
- **Desired Vol.Frac.**: specify volumetric ratio of the optimized design to the initial model.
- **Save Intermediate?**: save intermediate designs as STL for future review.
- **Optimize**: solve the optimization problem.
- **STOP OPTIMIZATION**: stop current optimization at current step.

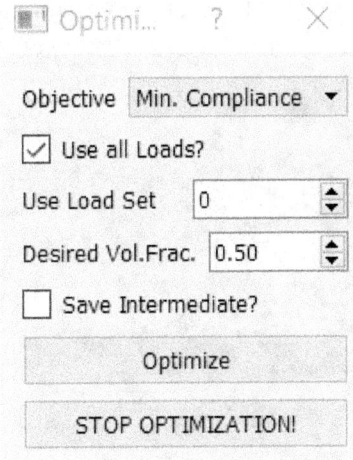

Figure 15.11: Optimization menu.

15.10 TopOpt Results

Once the optimization problem is solved, we can review the optimization results:
- **Topology@Vol**: display the intermediate topology at the specified volume fraction.
- **Save topology**: save the intermediate design at the specified volume fraction.
- **Compliance Pareto Curve**: plot compliance versus volume fraction.
- **Stress Pareto Curve**: plot maximum stress versus volume fraction.
- **Modal Pareto Curve**: plot the first eigenvalue versus volume fraction.

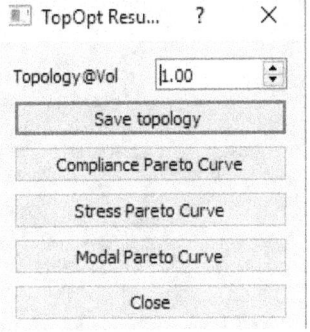

Figure 15.12: Results menu.

15.11 Lattice

Through *Lattice* menu, it is possible to generate cellular structures, where:
- **Lattice Type**: type of lattice structures.
- **Unit Size (mm)**: size of each cell.
- **Edge Division**: number of finite elements per cell edge.
- **Fill Ratio**: volumetric ratio of each cell structure to its corresponding solid structure.
- **Critical Surface**: retain which surfaces as solid.
- **Smoothening**: number of smoothening iterations.

- **Create Lattice**: generate the cellular lattice structure.
- **STOP DESIGN**: terminate current lattice design generation.

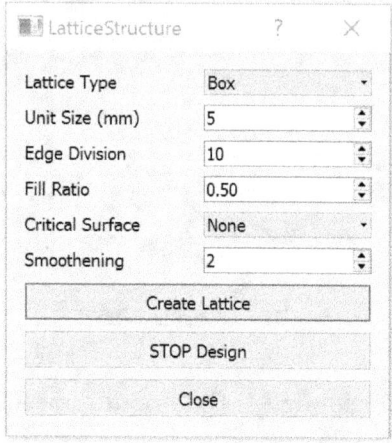

Figure 15.13: Lattice menu.

15.12 Projects

You can also save the current project through *Save Pareto Project* for your own future reference or for sharing with your colleagues. If you already have a project file with extension *.prj*, you can load it through *Load Pareto Project*.

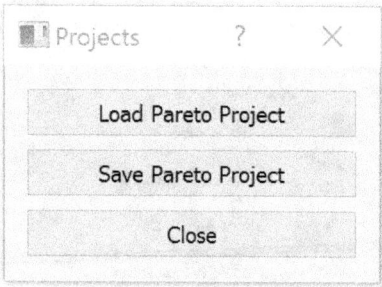

Figure 15.14: Projects menu.

www.ingramcontent.com/pod-product-compliance
Lightning Source LLC
Chambersburg PA
CBHW081116240526
45470CB00020B/3023